NUTR1R06i

NUTRIENT MANAGEMENT: REGULATORY APPROACHES TO PROTECT WATER QUALITY

VOLUME 1 – REVIEW OF EXISTING PRACTICES

David L. Clark
George Hunt
Michael S. Kasch
Paula J. Lemonds
Greg M. Moen
J.B. Neethling
HDR Engineering, Inc.

2010

WERF

The Water Environment Research Foundation, a not-for-profit organization, funds and manages water quality research for its subscribers through a diverse public-private partnership between municipal utilities, corporations, academia, industry, and the federal government. WERF subscribers include municipal and regional water and wastewater utilities, industrial corporations, environmental engineering firms, and others that share a commitment to cost-effective water quality solutions. WERF is dedicated to advancing science and technology addressing water quality issues as they impact water resources, the atmosphere, the lands, and quality of life.

For more information, contact:
Water Environment Research Foundation
635 Slaters Lane, Suite G-110
Alexandria, VA 22314-1177
Tel: (571) 384-2100
Fax: (703) 299-0742
www.werf.org
werf@werf.org

This report was co-published by the following organization.

IWA Publishing
Alliance House, 12 Caxton Street
London SW1H 0QS, United Kingdom
Tel: +44 (0) 20 7654 5500
Fax: +44 (0) 20 7654 5555
www.iwapublishing.com
publications@iwap.co.uk

Library of Congress Catalog Card Number: 2010925715

IWAP ISBN: 978-1-84339-617-8/1-84339-617-3

HDR Engineering, Inc.

The research on which this report is based was developed, in part, by the United States Environmental Protection Agency (EPA) through Cooperative Agreement No. CR-83155901-2 with the Water Environment Research Foundation (WERF). However, the views expressed in this document are not necessarily those of the EPA and EPA does not endorse any products or commercial services mentioned in this publication. This report is a publication of WERF, not EPA. Funds awarded under the Cooperative Agreement cited above were not used for editorial services, reproduction, printing, or distribution.

ACKNOWLEDGMENTS

The project team would like to acknowledge and thank the wastewater utilities involved in nutrient total maximum daily loads and NPDES permitting whose experiences were used as references in preparation of this report, including the Alexandria Sanitation Authority, VA, Arapahoe County Water and Wastewater Authority, CO, City of Coeur d'Alene, ID, Clean Water Services of Washington County, OR, Hampton Roads Sanitation District, VA, Hillsborough County, FL, Las Vegas, NV, LOTT Alliance, WA, Missoula, MT, Reno and Sparks, NV, Spokane County, WA, Stamford , CT, Village of Stamford, NY, and Village of Walton, NY. In particular, appreciation and thanks to the technical peer reviewers Robbin Finch, Will Hunley, and Kenneth Wood for providing detailed review comments and many supplemental references. Especially helpful were the many comments, suggestions and recommendations made by Dr. Amit Pramanik of WERF which greatly enhanced the value of this report.

Report Preparation

Principal Investigator:
David L. Clark, P.E., BSCE, MSCE
HDR Engineering, Inc.

Project Team:
George Hunt, P.E., Ph.D.
Michael S. Kasch, P.H., P.E, BSCE, MECE
Paula J. Lemonds, P.G., P.E., BSG, MSGE
Greg Moen, P.E., BSCE, MSCE
J.B. Neethling, P.E., Ph.D., BCEE
HDR Engineering, Inc.

Technical Review Committee

Robbin W. Finch
Boise City Public Works

Will Hunley
Hampton Roads Sanitation District (HRSD)

Kenneth N. Wood
Dupont

WERF Nutrient Removal Challenge Issue Area Team

James L. Barnard, Ph.D., D.Ing.h.c., Pr.Eng. BCEE
Terry L. Johnson, Ph.D., P.E., BCEE
Black & Veatch Corporation

Robbin W. Finch
Boise City Public Works

James A. Hanlon
James Wheeler, P.E.
U.S. Environmental Protection Agency

Joseph A. Husband, P.E., BCEE
Malcolm Pirnie Inc.

David Jenkins, Ph.D.
University of California – Berkeley

Gary R. Johnson, P.E., BCEE
Environmental Operating Solutions, Inc.

Michael McGrath, P.E.
Fairfax County Virginia

Sudhir Murthy, Ph.D., P.E.
District of Columbia Water & Sewer Authority

Tung Nguyen
Sydney Water Corporation

Rao Surampalli, Ph.D., P.E., DEE
U.S. Environmental Protection Agency – Region 7

G. David Waltrip, P.E.
Hampton Roads Sanitation District

Kenneth N. Wood, P.E.
DuPont

Heng Zhang, Ph.D., P.E.
Metropolitan Water Reclamation District of Greater Chicago

Water Environment Research Foundation Staff

Director of Research: Daniel M. Woltering, Ph.D.
Senior Program Director: Amit Pramanik, Ph.D., BCEEM

ABSTRACT AND BENEFITS

Abstract:

Utilities work with regulators to treat wastewater to levels that protect human health and ecosystems. Water quality criteria and permits are based on scientifically defensible and shared understanding of sources of pollutants in a watershed, as well as treatment capabilities and costs to control these in the aquatic environment. The national discussion of nutrient impacts on water quality continues to evolve – issues in high visibility water bodies such as the Chesapeake Bay, Long Island Sound, Gulf of Mexico, and Puget Sound highlight this. EPA's efforts to promulgate numeric nutrient standards in all states raise questions about how these standards apply to wastewater dischargers, whether they are effective, and how they affect WERF Subscribers and others in the water quality arena. This white paper report, *Nutrient Management: Regulatory Approaches to Protect Water Quality, Volume 1 Review of Existing Practices* (NUTR1R06i) provides a state-of-the art discussion of key nutrient management issues that confront point source wastewater dischargers nationwide. It provides a better understanding of challenges that utilities and regulators face setting and meeting low nutrient effluent limits, and it expands understanding of the practical capabilities of treatment technology.

Benefits:

- Presents a state-of-the art discussion of current topics associated with evolving numeric nutrient criteria as it relates to the capabilities of nutrient removal technologies and appropriate effluent discharge permit requirements.
- Provides background and context to current discussions and legal challenges in various parts of the United States with regard to receiving water quality impacts from nutrients.
- Discusses historical and emerging regulatory drivers and mechanisms, including watershed-based approaches, designed to help protect water quality in various eco-regions, states, and nationwide.
- Presents an overview of the capabilities and limits of advanced wastewater treatment technologies currently used for nitrogen and phosphorus removal, as well as unique considerations related to permitting to low nutrient levels. At low effluent limits, some portion of the remaining nitrogen and phosphorus in treatment plant discharges may not be removable with current treatment technology.
- Provides preliminary results from ongoing research to help define why nitrogen and phosphorus speciation is important, both in terms of biodegradability in wastewater treatment and bioavailability in the water environment.
- Reinforces previous findings that impairment in most watersheds is caused by a combination of point and nonpoint sources, or is dominated by nonpoint sources. Without nonpoint nutrient controls, technology based nutrient standards for wastewater discharges would have limited benefit for waterbodies nationally, and only combined point and nonpoint source nutrient management will be fully protective of the aquatic environment.

Keywords: Nutrients, nitrogen, phosphorus, nutrient removal, Clean Water Act, effluent limits, nutrient criteria, numeric nutrient standards, NPDES permit, total maximum daily load, TMDL, trading, variance, water quality.

TABLE OF CONTENTS

LIST OF TABLES

LIST OF FIGURES

LIST OF ACRONYMS

7Q10	Average Flow of 7 Consecutive Days with a 10 Year Return Frequency
30Q10	Average Flow of 30 Consecutive Days with a 10 Year Return Frequency
A^2O	Anaerobic-Anoxic-Oxic
AIZ	Allocated Impact Zone
ASIWPCA	Association of State and Interstate Water Pollution Control Administrators
AWWTF	Advanced Wastewater Treatment Facility
bEDNRP	Bioavailable Effluent Dissolved Non-reactive Phosphorus
bEDON	Bioavailable Effluent Dissolved Organic Nitrogen
BMP	Best Management Practice
BNR	Biological Nutrient Removal
BOD	Biochemical Oxygen Demand
btDON	Biodegradable Treatment Process Dissolved Organic Nitrogen
CBP	Chesapeake Bay Program
CDPHE	Colorado Department of Public Health & Environment
CFD	Computational Fluid Dynamics
COD	Chemical Oxygen Demand
CTIC	Conservation Technology Information Center
CWA	Clean Water Act
DO	Dissolved Oxygen
DON	Dissolved Organic Nitrogen
DOP	Dissolved Organic Phosphorus
DSSAM III	Dynamic Stream Simulation and Assessment Model
EBPR	Enhanced Biological Phosphorus Removal
EDNRP	Effluent Dissolved Non-reactive Phosphorus
EDON	Effluent Dissolved Organic Nitrogen
ENR	Enhanced Nutrient Removal
ENRP	Effluent Non-reactive Phosphorus
EON	Effluent Organic Nitrogen
EPA	Environmental Protection Agency
EPNRP	Effluent Particulate Non-reactive Phosphorus
EPON	Effluent Particulate Organic Nitrogen
iDNRP	Influent Dissolved Non-reactive Phosphorus

iDON	Influent Dissolved Organic Nitrogen
LA	Load Allocation
LMZ	Legal Mixing Zone
LOT	Limit of Technology
MF	Microfiltration
MGD	Million Gallons per Day
MOP	Manual of Practice
N	Nitrogen
NACWA	National Association of Clean Water Agencies
NF	Nanofiltration
NITG	State-EPA Nutrient Innovation Task Group
NOx	Nitrous Oxide
NPDES	National Pollutant Discharge Elimination System
NPS	Nonpoint Source
NRDC	Natural Resources Defense Council
P	Phosphorus
PAOs	Phosphorus Accumulating Organisms
PASS	Poly-Aluminum-Silicate-Sulfate
POTW	Publicly Owned Treatment Works
PPRE	Policy for the Potomac River Embayments
PS	Point Source
QUAL-2E	Enhanced Stream Water Quality Model
RDON	Refractory Dissolved Organic Nitrogen
RDOP	Refractory Dissolved Organic Phosphorus
rEDNRP	Recalcitrant Effluent Dissolved Non-reactive Phosphorus
rEDON	Recalcitrant Effluent Dissolved Organic Nitrogen
RO	Reverse Osmosis
rtDNRP	Recalcitrant Treatment Process Dissolved Non-reactive Phosphorus
rtDON	Recalcitrant Treatment Process Dissolved Organic Nitrogen
tDNRP	Treatment Process Dissolved Non-reactive Phosphorus
tDON	Treatment Process Dissolved Organic Nitrogen
TIN	Total Inorganic Nitrogen
SAB	Science Advisory Board
SBNR	Simultaneous Biological Nutrient Removal
SEM	Structural Equation Modeling
SPARROW	SPAtially Reference Regressions on Watershed model
SRT	Solids Retention Time

SSNC	Site-specific Nutrient Criteria
TDZ	Toxic Dilution Zone
TMDL	Total Maximum Daily Load
TN	Total Nitrogen
TP	Total Phosphorus
TPS	Technology Performance Statistic
TSD	Technical Support Document
UAA	Use Attainability Analysis
UCT	University of Cape Town
USGS	United States Geological Survey
VFAs	Volatile Fatty Acids
VNRP	Voluntary Nutrient Reduction Program
WESTCAS	Western Coalition of Arid States
WEF	Water Environment Federation
WEFTEC	Water Environment Federation Technical Exhibition and Conference
WERF	Water Environment Research Foundation
WLA	Waste Load Allocation
WQBEL	Water Quality Based Effluent Limit
WWTP	Wastewater Treatment Plant
ZID	Zone of Initial Dilution

EXECUTIVE SUMMARY

The purpose of this white paper is to provide a state-of-the art discussion of topics associated with evolving numeric nutrient criteria as it relates to the capabilities of nutrient removal technology and appropriate effluent discharge permit requirements. Key issues include the challenges for wastewater utility managers, private industry and permitting authorities in formulating appropriate nutrient limits, as well as compliance with these restrictive effluent standards, expanding the understanding of the capabilities of treatment technology, and the relationship between effluent limits and receiving water quality impacts.

Chapter 1.0 presents an introduction to key nutrient management issues that confront point source wastewater dischargers. The current United States Environmental Protection Agency (EPA) effort to promulgate numeric nutrient standards in all states has raised a number of questions about how these standards will be applied to wastewater dischargers and how they may impact Water Environment Research Foundation (WERF) members. The EPA effort that was initiated in 1998 with a goal of adopting numeric nutrient standards in all states by 2002 has not been accomplished.

The national discussion of nutrient impacts on water quality is evolving and is highlighted by issues in high visibility waterbodies such as Chesapeake Bay, Long Island Sound, the Gulf of Mexico, and Puget Sound. New voices have entered the discussion and some have called for treatment technology standards for nitrogen and phosphorus. For example, the Natural Resources Defense Council (NRDC) has recently filed a petition with EPA for rulemaking and called on EPA to add nitrogen and phosphorus limits to the treatment technology definition for secondary treatment.

In Chapter 2.0, receiving water quality issues are explored in greater detail, including nutrient criteria and key regulatory requirements that influence nutrient removal requirements. Targeted nutrient levels in lakes, streams, and estuaries can be very low concentrations that are challenging to meet with treatment of point sources and application of best management practices (BMPs) to nonpoint sources.

Chapter 2.0 presents a discussion of regulatory efforts to develop numeric nutrient standards, including key technical issues and a summary of progress in some key states. A number of nutrient related legal issues have resulted in interesting developments which may influence the approach to nutrient management in other areas of the nation.

Chapter 3.0 addresses key nutrient control challenges for dischargers. Waterbodies that receive effluent discharges have impoundments and reservoirs, irrigation diversions and returns, water supply withdrawals, and many other modifications that alter the aquatic environment. These conditions present challenging circumstances for the selection of appropriate in-stream nutrient targets that protect water quality and reflect realistically attainable conditions.

Nutrient removal treatment can substantially reduce point source discharges of nitrogen and phosphorus, however substantial investments are required to build and operate advanced wastewater treatment facilities. In some watersheds, nonpoint source nutrient loadings outweigh point sources to a degree that advanced treatment for nutrient removal, and even complete elimination of point sources, would have limited benefit to water quality. Nevertheless, point source NPDES permitted dischargers are the most directly regulated sources subject to nutrient control requirements resulting from numeric nutrient standards, total maximum daily loads (TMDLs), and water quality based permit limits. The costs of advanced wastewater treatment are substantial and Chapter 3.0 summarizes some of the cost and sustainability issues that should be balanced in order to make optimal decisions for nutrient management in watersheds.

Chapter 4.0 presents an overview of advanced wastewater treatment technology for nitrogen and phosphorus removal. Primary and secondary treatment processes only remove a limited fraction of nutrients from wastewater – a portion of the insoluble nitrogen and phosphorus taken out with primary solids and nutrient uptake required for biological growth. Nutrient removal also requires additional energy, chemicals, maintenance materials, and labor which increase the complexity of plant operations and costs.

The current state-of-the-art for nutrient removal is summarized in Chapter 4.0 and the capabilities of treatment technology are described. At low effluent limits, some portion of the remaining nitrogen and phosphorus in treatment plant effluent may not be removable with current treatment technology. Nitrogen and phosphorus speciation is an important area of nutrient research, both in terms of biodegradability in wastewater treatment and bioavailability in the water environment.

Chapter 5.0 presents a discussion of nutrient discharge permitting issues and some of the special considerations associated with appropriate limits for nitrogen and phosphorus. Surface water nutrient discharges should receive special considerations in discharge permitting for distinction from other effluent parameters, in particular toxic parameters, upon which much of the existing EPA permit writer's guidance is based. Appropriate NPDES discharge permit structures for nutrients should be based on long averaging periods, such as seasonal limits based on mean or median statistics. It is important that consideration be given to variability and reliability of effluent performance from advanced nutrient removal facilities.

Chapter 6.0 presents a discussion of nutrient discharge permitting issues and some of the special considerations associated with appropriate limits for nitrogen and phosphorus. Example discharge permits with nutrient limits are summarized for reference use. Since special considerations are required for appropriate surface water nutrient discharge limits, the summaries presented in Chapter 6.0 illustrate the range of effluent limits and variety of permit structures in place for some key nutrient removal facilities.

CHAPTER 1.0

INTRODUCTION

Excessive nitrogen and phosphorus loadings to watersheds impact water quality by stimulating the growth of algae which may result in depletion of dissolved oxygen, shifts in pH, degradation of habitat, impairment of drinking water sources, and in some cases harmful algal blooms. According to the EPA, nearly every state has nutrient related pollution with impacts in over 80 estuaries/bays, and thousands of rivers, streams, and lakes. In particular, EPA cites the Gulf of Mexico and the Chesapeake Bay as examples of significant water quality impacts from 35 States that contribute to nutrient loadings.

Nutrient loadings from both point and nonpoint sources contribute to water quality impairments in the nation's waterways. Point source discharges from wastewater treatment plants can be a significant source of nitrogen and phosphorus in watersheds. Nonpoint sources contribute substantial amounts of nutrients from land use activities such as agriculture, forestry, and urban/suburban development.

Targeted nutrient levels in lakes, streams, and estuaries can be very low concentrations that are challenging to meet with treatment of point sources and application of best management practices (BMPs) to nonpoint sources. Nutrient removal treatment can substantially reduce point source discharges of nitrogen and phosphorus, however substantial investments are required to build and operate advanced wastewater treatment facilities. In some watersheds, nonpoint source nutrient loadings outweigh point sources to a degree that advanced treatment for nutrient removal, and even complete elimination of point sources, would have limited benefit to water quality. Nevertheless, point source NPDES permitted dischargers are the most directly regulated sources subject to nutrient control requirements resulting from numeric nutrient standards, total maximum daily loads (TMDLs), and water quality based permit limits.

This chapter presents an introduction to key nutrient management issues that confront point source wastewater dischargers. In Chapter 2.0, receiving water quality issues are explored in greater detail, including nutrient criteria and key regulatory requirements that influence nutrient removal requirements. Chapter 3.0 addresses key nutrient control challenges for dischargers. The capabilities of wastewater treatment technology to control nutrient discharges is summarized in Chapter 4.0. Effluent discharge permitting issues are discussed in Chapter 5.0 and Chapter 6.0 presents example discharge permits with nutrient limits as references.

1.1 EPA 2007 Letter to States: Nutrient Pollution and Numeric Water Quality Standards

On May 25, 2007, Ben Grumbles, EPA Assistant Administrator, issued a memorandum entitled "Nutrient Pollution and Numeric Water Quality Standards" to State and Tribal water program directors. The memo provides an update on the EPA's commitment to accelerating the development of numeric nutrient water quality standards. The EPA published a national nutrient criteria strategy in 1998 and some States and Territories have made notable progress since then. As the EPA Assistant Administrator's memorandum notes, the overall progress has been uneven among States and Tribes over the past nine years.

Regarding the EPA's national nutrient strategy, the EPA noted that the focus is on the numeric nutrient criteria under the Clean Water Act and PA's strategy is to adopt numeric criteria to eliminate the need to translate narrative criteria.

The EPA has four focus areas for nutrient criteria, including:

1. Working with states that are further away from adopting numeric criteria through the use of training and workshops;
2. Providing direct assistance to States that are close to adopting numeric N and P criteria;
3. Developing science-based approach for developing new Section 304(a) criteria for estuaries, wetlands, and large rivers; and
4. Clearly and effectively communicating the dangers of nutrient pollution and the merits of numeric nutrient criteria to states, nutrient sources, and the general public.

Grumbles noted that high nitrogen and phosphorus loadings result in harmful algal blooms, reduced spawning grounds and nursery habitats, fish kills, oxygen-starved hypoxic or "dead" zones, public health concerns related to impaired drinking water sources, and increased exposure to toxic microbes, such as cyanobacteria. The most widely known examples of significant nutrient impacts include the Gulf of Mexico and the Chesapeake Bay.

Descriptions of four watersheds of particular importance are included in the memo:

Chesapeake Bay has an existing hypoxia problem with population increasing in the watershed by 150,000 each year.

The Gulf of Mexico has a prevalent and well-documented hypoxic "dead" zone. Thirty-one states contribute to the watershed and, through the 2001 Hypoxia Action Plan, the EPA Science Advisory Board reports that phosphorus plays a much greater role in the hypoxia problem than previously thought.

In the Long Island Sound, DO is below standards in one-third to one-half of the Sound. Nitrogen loadings have been capped at 1990 loads, and a water quality trading program has been implemented in Connecticut for point sources with a market-based approach.

In Puget Sound, the highest priority is to gain a better understanding of nutrient and bacteria loadings from septic systems through the Puget Sound Action Plan.

1.2 Association of State and Interstate Water Pollution Control Administrators

In the EPA memorandum from Assistant Administrator Ben Grumbles, State environmental agencies were encouraged to accelerate the pace of the development of numeric nutrient water quality standards (EPA, 2007). The request for acceleration also included an offer from the EPA Office of Water to assist those States close to adopting numeric criteria and building capacity with those States further from adopting numeric criteria. Additionally, EPA sought to build a science-based foundation for developing criteria for estuaries, wetlands, and large rivers. Finally, EPA placed additional emphasis on the nutrient criteria and continues to work to communicate to States and the public the dangers of nitrogen and phosphorus pollution and the merits of numeric nutrient criteria.

Reactions and responses to Ben Grumbles memorandum included letters from the Association of State and Interstate Water Pollution Control Administrators (ASIWPCA). ASIWPCA's first response was a letter directly to Ben Grumbles on July 18, 2007 (ASIWPCA, 2007a). Overall, ASIWPCA seemed to applaud EPA's initiative to accelerate the process and provide more definition for nutrient controls. "Member States agree that nutrient controls are critical and necessary component of comprehensive water quality management" (ASIWPCA, 2007a). ASIWPCA also urged EPA to accelerate promulgation of numeric water quality criteria for nutrients with the caveat that EPA "should develop categorical standards for POTWs and have consistent realistic effluent limits" (ASIWPCA, 2007a).

While ASIWPCA appeared to be supportive of the EPA initiative, they also expressed some concerns including seven items explicitly highlighted in the letter. These concerns include the need for more science, supporting State nutrient criteria development plans, appropriate implementation, consideration of impacts on NPDES permits, quantitative economic assessment, categorical standards, and accurate portrayal of States progress on the issue. ASIWPCA expressed concern about the development of generic or regional criteria. Concerns included "The uniformity of eutrophic and productivity conditions that numeric criteria would promote defies both common sense and basic principles of ecological succession" and "because no two waterbodies are the same, site-specific evaluations and, most probably, site-specific criteria are required that reflect their uniqueness and protect their natural trophic tendencies" (ASIWPCA, 2007a). ASIWPCA emphasized the need to establish effluent guidelines, or similar guidance with realistic expectations for implementation, that addresses more than numerical criteria but also address how, when, and where to apply nutrient standards to improve and protect water quality.

ASIWPCA also pointed out "that nutrient pollution in many watersheds is largely or exclusively attributed to non-point sources" (ASIWPCA, 2007a). States need help in recognizing all impacts to water quality and developing a comprehensive approach to reducing nutrients. Local changes will be necessary but will likely result in few changes locally but hopefully contribute to the water quality of the entire watershed. Large watersheds like the Chesapeake and Gulf of Mexico will need improvement far upstream in the Midwest.

ASIWPCA's second response was a letter to Ephraim King, Director of Office of Science and Technology, Office of Water, EPA on November 26, 2007 (ASIWPCA, 2007b). ASIWPCA sought a Federal partnership with States on treatment technology. The challenges of addressing nutrients in multi-state watersheds would benefit from a technology-based approach to provide consistency. ASIWPCA encouraged EPA "to research the concept of best available technology or reasonable treatment technologies as a first step towards controlling" nutrients (ASIWPCA, 2007b). ASIWPCA argued that "state-by-state, permit-by-permit, would be more difficult, lengthier and will result in less consistent controls than setting federal technology-based standards" (ASIWPCA, 2007b). ASIWPCA stated that the consistent application of technology-based standards has many benefits.

In 2008, ASIWPCA published a document entitled "Call for Change" which included a section with the heading "An Effective Strategy for Dealing with Nutrients" (ASIWPCA, 2008). Again ASIWPCA stated their position on the need for a holistic national strategy on nutrient pollution. The need for an effective national nutrient strategy across the board from all Federal and State agencies that deal with water resources was recommended. Other recommendation also echoed previous statements, including the need from improvements from both point and non-point sources. The need for economic considerations and cost-effectiveness evaluations was recommended as part of the strategy. The recommendation for "a sustainable technology-based approach for point sources and a regulatory/technology-based approach (or some equally effective strategy) for nonpoint sources" was also reiterated (ASIWPCA, 2008). ASIWPCA stance is that a regulatory based approach is the most efficient and rapid path forward, as opposed to the case-by-case water quality based approach to achieving nutrient reductions for individual waterbodies.

1.3 Effluent Guidelines

As defined by the EPA, effluent limitations guidelines are national technology-based standards that are developed by the EPA for specific industries and manufacturing processes and are intended to represent the best practicable, conventional or available pollutant reductions that are economically achievable for an industry (EPA, 2008). Per Section 304(b) and (m) of the Clean Water Act (CWA), the EPA is required each year to generally review its effluent limitations guidelines regulations and every other year, to publish a plan that establishes a schedule for detailed review and possible revision of existing standards and other industries discharging toxic or "nonconventional" pollutants for which effluent guidelines have not been defined.

In 2008, the EPA released its Final 2008/2009 Effluent Guidelines Program Plan. The EPA reviewed the existing effluent guidelines and pretreatment standards, and evaluated point sources and indirect dischargers that do not have categorical standards to identify potential new categories. The biennial Effluent Guidelines Program Plan, which is required under CWA Section 304(m), identifies any new or existing industrial categories selected for effluent guidelines rules and provides a schedule for the establishment of these rules.

1.4 NRDC Petition for Rulemaking on Secondary Treatment

In November 2007, the Natural Resources Defense Council (NRDC) filed a Petition for Rulemaking with the EPA to limit nutrient pollution from wastewater treatment facilities. Ten other regional and national environmental groups, including the Sierra Club and American Rivers, joined NRDC in the petition. NRDC argues that nitrogen and phosphorus effluent limitations should be a part of the base technology definition of secondary treatment. NRDC contended that the EPA must protect the public by establishing nutrient limits, specifically that the EPA unreasonably delayed publishing information on Secondary Treatment to remove excess nutrients. The NRDC also notes that nutrient control is properly included within "Secondary Treatment" and cites the following effluent nutrient levels as attainable:

- Effluent TP 0.3 mg/l and TN 3 mg/l are Consistently Attainable Using Current Technology
- Effluent TP 1 mg/l and TN 8.0 mg/l is Attainable with Existing Technology Using Only Improved Biological Treatment Processes

NRDC considers the EPA's approach to site-specific nutrient standards as unreasonable and that the EPA cannot rely on a water-quality based approach to control nutrient pollution. NRDC argues that nutrient pollution is widespread and justifies a generally-applicable standards approach to treatment for nutrients. NRDC cites the EPA as evidence of this conflict:

"Nutrient pollution is widespread. The most widely known examples of significant nutrient impacts include the Gulf of Mexico and the Chesapeake Bay. For these two areas alone, there are 35 States that contribute the nutrient loadings. There are also known impacts in over 80 estuaries/bays, and thousands of rivers, streams, and lakes. Virtually every State and Territory is impacted by nutrient-related degradation of our waterways. All but one State and two Territories have Clean Water Act Section 303(d) listed impairments for nutrient pollution. States have listed over 10,000 nutrient and nutrient-related impairment. Fifteen States have more than 200 nutrient-related listings each." (EPA, 2007).

NRDC regards the EPA's reliance on site-specific standards as unreasonable given the pervasive nutrient pollution and the lack of numeric nutrient standards, which hinder the ability to require water-quality based effluent limitations. NRDC calls for the EPA to specify the degree of nitrogen and phosphorus reduction attainable through secondary treatment.

1.5 National Association of Clean Water Agencies on the NRDC Petition

The National Association of Clean Water Agencies (NACWA) provided comments on the NRDC Petition for Rulemaking on Secondary Treatment in a letter to EPA dated February 29, 2008 (NACWA, 2008). NACWA expressed concern that the NRDC petition calling for nutrient limits as part of the secondary treatment process is not technically or financially practical, and that the approach is not the most effective or environmentally sensitive way to reduce nutrient pollution. NACWA criticized the proposed "one size fits all" approach to a water quality problem that is site-specific and best addressed through site-specific measures.

NACWA believes site-specific water quality efforts will be more effective than technology-based nutrient removal limits and cites five areas of concern with the NRDC Petition:

(1) Legal basis for incorporating nutrient removal into secondary treatment
(2) Failure of the petition to address the contribution to nutrient loadings from non-point sources
(3) Potentially high costs for treatment plants to meet a national nutrient limit and whether such expenditures are cost-effective
(4) Increased negative environmental impacts of mandating a national nutrient removal limit
(5) Inappropriateness of national limits for local and regional water quality issues

NACWA points out that tertiary treatment for nutrient removal goes well beyond the original Congressional intent in the Clean Water Act for secondary treatment and that EPA has denied previous requests to include nutrient removal as part of secondary treatment standards. NACWA notes that the petition fails to acknowledge the impact of nonpoint sources of nutrients on water quality. Reliance on wastewater treatment technology alone will not be as effective in improving water quality as a water quality-based approach to controlling both point and nonpoint sources. Water quality-based projects are technically achievable and provide the best combination of environmental benefits, cost-effectiveness, and local flexibility to manage nutrients.

NACWA members have found that the costs for nutrient removal are substantial and that major expenditures are associated with expanding facilities because of the need for more plant site and infrastructure to accomplish necessary retrofits and upgrades.

There are a number of unintended negative environmental consequences of requiring across-the-board nutrient removal in terms of carbon footprint and increased quantities of biosolids for disposal. Nutrient removal treatment requires substantial electrical power, resulting indirect greenhouse gas emissions, as well as direct emissions of greenhouse gases and nitrous oxide emissions, in particular. The adverse impacts of point source nutrient controls should be balanced with consideration of nonpoint source controls, which consume little energy and may reduce greenhouse gas emissions by sequestering carbon.

NACWA notes that a "one size fits all" approach to nutrient effluent limits is inappropriate because of the site-specific effects of nutrients and variability from watershed to watershed. The assimilative capacity of watersheds varies, as does the importance of nitrogen and/or phosphorus in different waterbodies, depending upon site-specific conditions. NACWA endorses the appropriate determination of site-specific water quality limits based on the established mechanisms in the Clean Water Act that allow communities to respond appropriately to the environmental needs of local water bodies in a cost effective manner.

1.6 References

ASIWPCA. 2007a. Nutrient Pollution and Numeric Water Quality Standards Memo. Association of State and Interstate Water Pollution Control Administrators, Washington, D.C.

ASIWPCA. 2007b. Technology Thresholds for Nutrients. Association of State and Interstate Water Pollution Control Administrators, Washington, D.C.

ASIWPCA. 2008. Call for Change. Association of State and Interstate Water Pollution Control Administrators, Washington, D.C.

EPA, 2008. Accessed on: November 5, 2008. http://www.epa.gov/guide/questions/.

EPA. 2007. Memorandum Nutrient Pollution and Numeric Water Quality Standards. From Ben Grumbles, EPA Assistant Administrator for Water, to State Water Program Directors et al. Washington, D.C.

National Association of Clean Water Agencies (NACWA), Letter to Benjamin Grumbles, Assistant Administrator, Office of Water, U.S. EPA, regarding NRDC Petition for Rulemaking on Secondary Treatment, February 29, 2008.

CHAPTER 2.0

RECEIVING WATER QUALITY DRIVERS

Nutrient control requirements can be challenging to meet because the nitrogen and phosphorus concentrations that may lead to water quality impairment may be very low concentrations. This watershed sensitivity to nutrient enrichment results in a need for careful consideration of nutrient targets for water quality protection and an understanding of the need for management of both both point and nonpoint contributions.

This chapter presents receiving water quality issues, including nutrient criteria and key regulatory requirements that influence nutrient removal requirements. Nutrient water quality drivers including TMDLs, narrative criteria, numeric criteria and other methods of nutrient target setting are discussed. Water quality based effluent limits, use attainability, water quality variances, anti-degradation requirements, and adaptive management are summarized as they relate to nutrient discharge limitations. Chapter 4.0 follows with an overview of wastewater treatment technology to control nutrient discharges. Effluent discharge permitting issues are discussed in Chapter 5.0 and Chapter 6.0 presents example discharge permits with nutrient limits as references.

2.1 Total Maximum Daily Load (TMDL) Process

Water quality limited receiving waters (CWA Section 303(d) lists), total maximum daily loads (TMDLs), and the wasteload allocations established under this process are leading to new challenges for wastewater treatment plants to control nutrient discharges. In many States, federal district court rulings on TMDL lawsuits in the 1990s resulted in requirements for State environmental agencies and the EPA to accomplish TMDLs on all Section 303(d) listed water quality limited stream segments on accelerated schedules. TMDLs and wasteload allocations are leading to wastewater treatment plant discharge limitations for nutrients and these limitations provide the basis for new discharge permits that require upgrades to advanced treatment. The water quality studies required to develop TMDLs can be challenging, especially the determination of nutrient targets and the development of appropriate effluent discharge permit structures for wastewater treatment plants.

EPA describes the TMDL process as a successive progression of steps that yield a TMDL as follows:

- Problem identification
- Identification of water quality indicators and targets
- Source assessment
- Linkage between water quality targets and sources
- Allocations

- Follow-up monitoring and evaluation
- Assembling the TMDL

This framework is to be completed concurrently, or iteratively, to produce a legally approvable TMDL with load calculations and allocations, which support the basis for review by EPA. However, in actual practice, the process is far from straightforward. Many waterbodies have been altered far beyond natural conditions. Assessing the complexities of watersheds and meeting the challenges in establishing appropriate target conditions to address water quality impairments is difficult. Incomplete water quality data compounds the challenge of formulating TMDLs, as it results in an incomplete understanding of all of the point and nonpoint source loadings that result in the impaired condition. Frequently, the TMDL process is undertaken without the key stakeholders responsible for the point and nonpoint source loadings having an adequate understanding of the potential impacts of the load allocations that may result from the process. This can lead to situations where the TMDL may be reviewed and approved by EPA, but cannot practically or economically be implemented.

EPA expects that public participation for Section 303(d) activities (impaired waters lists and TMDL development) must be consistent with Section 101(e) of the CWA, which requires EPA and States to ensure public participation "*in the development, revision, and enforcement of any regulation, standard, effluent limitation, plan, or program established ... under the Act.*" Specifically, EPA regulations require States to provide ample opportunity for public participation in the development of lists of impaired waters and the development of TMDLs under Section 303(d). Public participation requirements are outlined in 40 CFR Part 25. In addition, Section 303(d)(2) (40 CFR 130.7(a)) provides that the process for developing Section 303(d) lists and public participation be described in the State Continuing Planning Process under Section 303(e).

Public participation is that part of the decision making process through which responsible officials and other stakeholders become aware of public attitudes by providing ample opportunity for interested and affected parties to communicate their views. Public participation includes providing access to the decision making process, seeking input from and communicating with the public, assimilating public viewpoints, and preferences and demonstrating that those viewpoints and preferences have been considered by the decision making official.

2.1.1 Nutrient TMDLs

Nuisance aquatic growth driven by nutrient loadings can impair a waterbody's designated beneficial uses in a number of ways, including: interfering with recreational activities; creating aesthetic issues (e.g., odors, filamentous algal growth); covering substrate that provides habitat for aquatic organisms and fish reproduction; consuming dissolved oxygen as a result of decay; shifting pH and dissolved oxygen by algal respiration; and degrading water supplies (e.g. taste and odor impacts). The challenge in developing nutrient TMDLs is in selecting the targeted water quality conditions for receiving waters and the nitrogen and phosphorus concentrations associated with those conditions.

Narrative standards are currently the most common criteria for nutrients, because several factors in addition to nutrient concentrations determine the impact of nutrients on receiving water quality. The additional factors that influence the extent of algal growth include: light penetration; stream velocity and scour; frequency and intensity of flood events; substrate stability; grazing;

and temperature. For these reasons, nutrient concentrations that drive enriched conditions in one stream may not impair the beneficial uses in another.

Conducting the studies necessary to support nutrient TMDLs and select appropriate target water quality conditions can be time consuming and expensive. Where states face many impairment listings on many waterbodies, schedule demands and resource limitations constrain the effort to conduct detailed individual analysis to support nutrient TMDLs.

EPA recognizes that nutrient TMDLs can be technically challenging, and it is imperative to use all existing and readily available data and information when developing these TMDLs. EPA also recognizes the importance of assuring that both near-field and downstream impacts from nutrient loadings be considered in establishing wasteload allocations and load allocations. Without numeric criteria, EPA and the states generally rely on other intermediate water quality indicators such as dissolved oxygen, chlorophyll a, and water clarity to establish nutrient TMDL endpoints. Further, EPA advocates that the watershed approach to TMDL development be considered whenever possible and practical. This approach will allow for the consideration of loads and potential impacts throughout the entire watershed.

2.1.2 Incomplete Water Quality Data

A frequent problem in TMDL development is the lack of complete water quality data to allow a full understanding of all of the loadings impacting a watershed. The pursuit of additional water quality data is often an initial step in the TMDL process. However, data collection and analysis is time consuming and expensive – conditions that are often inconsistent with the demands of the production schedule and the resources available. Even with additional monitoring, a full definition of all loadings in a watershed is difficult to attain.

Water quality modeling efforts designed to provide a complete understanding of watershed conditions are especially data intensive and time consuming. As a result, the most sophisticated modeling tools – those with the potential for providing the fullest understanding of watershed functions – are often reserved for only the highest priority watersheds. Even with the use of the most sophisticated models, adequate data and acceptable calibration is a challenge because watersheds are so complex.

Point source discharge loading data is often the most readily available in terms of quantifiable pollutant loading information. This occurs by virtue of the CWA's National Pollutant Discharge Elimination System (NPDES) permit reporting requirements. Monthly discharge monitoring reports must be submitted to regulatory agencies for all point source dischargers in every watershed. Unfortunately, for point source dischargers, the availability of the data has sometimes been interpreted as an indication that point sources are the only loadings that need to be controlled in TMDLs. This is certainly not the case in most watersheds, especially with nutrients – where the impairment is caused by either nonpoint sources, or a combination of point and nonpoint sources. EPA's policy requires that all sources of the impairing pollutant be considered when establishing TMDLs, however the ability to control nonpoint sources of pollutants has been problematic because of the lack of required and/or enforceable mechanisms.

Some water quality monitoring data is generally available in TMDL watersheds since it is that data which provides the basis for the impairment designation. However, water quality data does not necessarily provide the information necessary to associate pollutant loadings with nonpoint sources such as agriculture, forestry, urban/suburban drainage, etc. Inappropriately estimating natural background loadings can be an especially problematic in the resulting TMDL. If all unidentified loadings are characterized as natural background, potentially manageable nonpoint source loadings may not be quantified and designated for reduction. Conversely, if natural background loadings are underestimated, TMDL load reductions may exceed what is possible.

Groundwater, both direct and indirect discharges, can also be an important component of overall watershed nutrient loadings. The lack of direct monitoring data, however, may disguise its importance. Land use activities such agriculture and forestry, as well as urban/suburban drainage and the use of on-site septic systems all may result in pollutant loadings to groundwater that is then tributary to surface waters. The groundwater/surface water interactions are typically complex and difficult to understand. However, in many important watersheds, groundwater delivery of nonpoint source nutrient loadings is very prevalent. Understanding this interconnection can lead to substantially different management activities in the watershed to comply with TMDL requirements.

2.1.3 Schedule and Resource Limitations

The time and resources necessary to develop a complete and scientifically well-founded TMDL for nitrogen and phosphorus can be substantial. Rarely are the time and resources adequate to satisfy those charged with the responsibility to prepare TMDLs for nutrients. Court ordered TMDL schedules compound the challenges by adding the pressure of mandated deadlines for completion. This can be quite frustrating to TMDL leaders seeking a complete scientific understanding of the watershed. While budgets and time may be limited, the scrutiny with which the TMDL will be reviewed is not. The potential for critical review to contest the water quality analysis and resulting loading allocations is real.

Since budget and time are limited, reducing the workload required to prepare a TMDL is attractive. Often, consideration of TMDL implementation is abbreviated because it is not a mandatory component of an acceptable TMDL for nutrients. This is unfortunate, since implementation planning presents an opportunity to engage both point and nonpoint stakeholders and reveal potentially impractical aspects of the TMDL. Implementation planning calls for the examination of water quality requirements and the TMDL process in a way that translates more directly to the actions that will be required for compliance. Concurrent nutrient TMDL development and implementation planning can result in more practical watershed management plans with greater stakeholder support.

2.1.4 Engagement and Communication with Key Stakeholders

Many TMDLs are developed without key stakeholders having the understanding necessary to accept the results of the TMDL and embrace the activities that may be required for watershed restoration. This seems to be an especially difficult problem to overcome and the misunderstandings that arise can compromise the efforts to improve water quality.

The TMDL process itself is complex enough that often only those preparing the water quality analysis understand who may be impacted and in what way. Often the State regulatory agency is leading the development of the TMDL and has the combined burden of conducting the analysis and communicating with the regulated community about the implications. Most in the regulated community are fully consumed with the core demands of their primary responsibilities and unfamiliar with the TMDL process. This disconnect can result in the development of technically impractical and unaffordable TMDLs for nitrogen and phosphorus.

Point source stakeholders and nonpoint source resource managers are generally the most skilled and experienced with managing the loadings from their sources. Wastewater utilities and private industry have the knowledge, skill and experience to understand the limits of their respective and applicable treatment technologies for nutrient removal. However, if these stakeholders not engaged in the TMDL process, they may be confronted with wasteload allocations that are not practically or economically attainable.

Similarly, for the nonpoint sources, nutrient load allocations may be formulated that call for reductions exceeding what can be accomplished with best management practices (BMPs) with reasonable assurance. Since BMP effectiveness is less deterministic than point source controls, meeting reasonable assurance requirements in a robust way can be challenging.

Without complete engagement of the point and nonpoint source stakeholders to provide a well-founded understanding of the effectiveness of nutrient reduction efforts, the potential application of some of the most innovative and economical watershed management approaches may be inadvertently limited. Water quality trading and loading offsets between point and nonpoint sources holds the promise of providing optimal watershed management plans. However, incomplete nutrient loading analysis, limitations of time, and the lack of complete engagement of both point and nonpoint source stakeholders limit the potential for these types of approaches to provide economical water quality improvements.

EPA advocates public involvement with interested stakeholders and supports a watershed scale TMDL approach. EPA believes that watershed permitting and the implementation of EPA 's trading policies have the potential to result in the most cost effective and optimal watershed management plans that are developed at the watershed wide scale.

2.2 Technology and Water Quality-Based Effluent Limitations

The Clean Water Act includes a prohibition against the discharge of pollutants to waters of the U.S. without a permit. The wastewater permit program in the US is known as the National Pollutant Discharge Elimination System (NPDES). Only the EPA or a state with permitting authority can issue an NPDES permit.

NPDES permits include effluent limits. These limits may restrict the discharge rate, concentration and/or quantities of pollutants authorized for discharge under the permit. The effluent limitations may be set with a consideration of average levels of pollutants, industrial standards for treatment technology, dilution, mixing zones and total maximum daily loads.

The Clean Water Act includes a two-tiered approach for effluent limits. The first tier is technology-based limits or the technological feasibility of achieving industry standards. The second tier is water quality-based limits or requirements based on the quality of the receiving water.

The permit writer develops technology-based effluent limits for regulated pollutants proposed for discharge. These minimum effluent limitations are determined based on the industry and available technology. The goal is commonly referred to as a "performance goal" and ultimately, "zero discharge to the maximum extent practicable." Technology-based effluent limitations for industrial sources are often developed based on the nationally applicable Effluent Guidelines and when these are non-existent, effluent limits are based on the exercise of best professional judgment. From the EPA Permit Writer's Manual (EPA, 1996):

> *There are two general approaches for developing technology-based effluent limits for industrial facilities: (1) using national effluent limitations guidelines (ELGs) and (2) using Best Professional Judgment (BPJ) on a case-by-case basis (in the absence of ELGs). Technology-based effluent limits for municipal facilities (POTWs) are derived from secondary treatment standards. The intent of a technology-based effluent limitation is to require a minimum level of treatment for industrial/municipal point sources based on currently available treatment technologies while allowing the discharger to use any available control technique to meet the limitations.*

Municipal dischargers are required to meet secondary treatment, pursuant to sections 301(b)(1)(B) and 304(d)(4) of the Clean Water Act and codified in the regulations at Part 133. From the EPA Permit Writer's Manual (EPA, 1996):

> *Similar to its approach for controlling the discharges from industrial sources, the 1972 CWA required POTWs to meet performance-based requirements based on available wastewater treatment technology. Section 301 of the CWA established a required performance level, referred to as "secondary treatment," that all POTWs were required to meet by July 1, 1977.*
>
> *More specifically, Section 301(b)(1)(B) of the CWA requires that EPA develop secondary treatment standards for POTWs as defined in Section 304(d)(1) of the Act. Based on this statutory requirement, EPA developed secondary treatment regulations which are specified in 40 CFR Part 133.*

The permit writer then compares the limitations to the specific waterbody, its water quality, applicable TMDLs, and the reasonable potential of the discharge to impair the water quality. If the technology effluent limits are not protective, then more stringent water quality-based effluent limits are added to the permit. These goals are commonly referred to as "fishable and swimmable," and "TMDL wasteload allocations." Wasteload allocations developed in TMDLs do not in themselves lead to NPDES permits. Permits need only be consistent with the assumptions and calculations developed in the TMDL. The final effluent limits will also need to be crafted to be consistent with the applicable water quality standards.

While the EPA has established guidelines on technology-based limits, water quality-based effluent limits can range considerably depending on the receiving water. For example, receiving waters with a beneficial use that sets high water quality standards and potentially has periods of low flow, dominated by the discharge, can result in extremely low limits. Such a situation can create a difficult challenge for the discharger.

Low effluent limits may force a discharger into a situation where they have to construct expensive treatment upgrades or find a new location for their discharge. An additional option for the water quality-based effluent limit is the potential for trading. An example of the type of permit language that includes trading is as follows:

"The Discharger must meet, through treatment or trading, a mass-based effluent limitation for Pollutant A of <insert baseline>. If this effluent limitation is met through trading, the Discharger must purchase credits from authorized Sellers in an amount sufficient to compensate for the discharge of Pollutant A from Outfall 001 in excess of <insert baseline>, but at no time shall the maximum mass discharge of Pollutant A during <insert averaging period> exceed the minimum control level of <insert minimum control level>. Thus, the maximum mass discharge of Pollutant A to be offset through credit purchases is <insert minimum control level – baseline>." (EPA, 2007a).

2.3 Use Attainability Analysis

A use attainability analysis (UAA) is an analysis that must be performed to assess and document the factors affecting attainability of the use and to determine the highest attainable use for a waterbody. If the highest attainable use differs from the current one in a state's regulations, then the use can be changed if one of the factors under 40 CFR 131.10(g) are met and if the state can demonstrate that attaining the designated use is not feasible because:

1. Naturally occurring pollutant concentrations prevent the attainment of the use.
2. Natural, ephemeral, intermittent or low flow conditions or water levels prevent the attainment of the use, unless these conditions may be compensated for by the discharge of sufficient volume of effluent discharges without violating State water conservation requirements to enable uses to be met.
3. Human caused conditions or sources of pollution prevent the attainment of the use and cannot be remedied or would cause more environmental damage to correct than to leave in place.
4. Dams, diversions, or other types of hydrologic modifications preclude the attainment of the use, and it is not feasible to restore the waterbody to its original condition or to operate such modification in a way that would result in the attainment of the use.
5. Physical conditions related to the natural features of the waterbody, such as the lack of a proper substrate, cover, flow, depth, pools, riffles, and the like, unrelated to water quality, preclude attainment of aquatic life protection uses.
6. Controls more stringent than those required by sections 301(b) and 306 of the Clean Water Act would result in substantial and widespread economic and social impact.

States determine waterbody designations based upon existing uses and the potential to meet the objectives of the Clean Water Act. The Clean Water Act required the recognition of existing uses, which are actual uses that have been attained in a waterbody on or after November 28, 1975, the date EPA promulgated the first water quality standards. Designated uses are to be set at the highest attainable use, which can reflect better water quality than needed for "fishable/swimmable". Additionally, states are to consider the use and value of the waterbody for public water supply; for protection and propagation of fisheries, aquatic life and wildlife; and for recreational, agricultural, industrial, and navigational purposes.

States may designate multiple uses for a waterbody and may split a waterbody into multiple sections with different uses. Federal law requires that the most sensitive of the beneficial uses be attained. Examples of beneficial use designations include public water supply, primary contact recreation, and aquatic life uses, such as cold and warm water habitat. The Clean Water Act concept of zero discharge, or no discharge of pollutants without a permit, means that waste transport or treatment is not an acceptable use.

The requirement that the states determine beneficial uses resulted in a variety of designations. Some states took the approach of selecting beneficial uses associated with higher water quality standards for a majority of their waterbodies. This resulted in many of the waters identified as impaired, requiring a 303(d) listing and TMDLs including nutrients. Some states took the approach of selecting beneficial uses associated with lower water quality standards for a majority of their waterbodies. This resulted in lawsuits claiming that the guidelines for determining beneficial uses were not followed.

In order to remove a beneficial use Federal law requires a UAA. The UAA must demonstrate that the beneficial use does not exist, cannot be attained, justify the lower level standard for the waterbody, and the UAA must demonstrate what the highest attainable use is. The UAA is a thorough and complicated scientific assessment of biological, chemical, physical, and economic conditions. Historically, it has been difficult to prove a current beneficial use to be not feasible and have the designation removed. States continue to work on guidance regarding the UAA.

WERF continues to develop technical and policy tools related to the UAA process. WERF has published a study entitled the "New Study on Factors for Success in Developing Use Attainability Analyses" that identifies factors for success in the UAA process, and suggests guidance for applying these factors. This study evaluates numerous UAAs, identifying key challenges and providing helpful information about how best to address the challenges. The case studies address three common situations in which UAAs are being considered nationwide: wet-weather impacts; urban settings; and effluent-dependent or -dominated streams (WERF, 2007).

2.4 Water Quality Variances

A water quality variance provides temporary relief from meeting water quality standards in order to avoid downgrading the designated use of the waterbody. However, the variance cannot further degrade the existing condition and does not exempt any technology-based effluent limitations. Additionally, for any variance, the underlying water quality standard remains in effect.

Variances are generally limited in scope to specific sources and pollutants and specific portions of the receiving water body. "*Not all states/tribes grant variances. In a 1990 assessment of state variance procedures, it was found that only 16 of 57 States/Territories had granted variances, and some of those had done so infrequently*" (EPA, 2010b and EPA, 2010c). Typically, a variance is for only one pollutant. Additionally, a variance may be limited in space and time. Since dischargers usually seek variances because of some uncertainty, variances include a range of limitations.

The categories of variances are single discharge, multiple discharges, variance for a whole waterbody, or variance as part of a restoration, acknowledging achievement of only partial attainment. Any one of the following six factors can demonstrate that the designated use is not feasible to attain in the short term (EPA, 2008):

1. Naturally occurring pollutant concentrations
2. Dams or other hydrologic modifications
3. Natural, ephemeral intermittent low-flow
4. Natural physical conditions preclude attainment of aquatic life uses
5. Human-caused conditions or pollutant sources that cannot be remedied or would cause more environmental damage to correct than to leave in place
6. Substantial and widespread economic and social impact

The most common of the six factors used to justify a variance in practice is "the substantial and widespread economic impact" (EPA, 2010a, EPA, 2010b). A water quality standard variance may not be approved if it places an additional burden on other point and non-point sources (USC, 2010, NYU, 2010). A variance will also be denied if it potentially harms a critical habitat or jeopardizes any threatened or endangered species (USC, 2010, NYU, 2010).

If a discharger receives a variance, the variance will typically not exceed the term of the NPDES permit (five years). The variance may include a schedule that demonstrates reasonable progress towards meeting the water quality standards. Other provisions may include additional monitoring and reporting. In granting the variance, the state must follow its established variance policies and the variance is subject to public and EPA review. The discharger may reapply to renew the variance, but must demonstrate that the receiving water beneficial use in question is still not attainable.

2.4.1 Example Nitrogen Variance

An example of a nitrogen variance is the variance granted to the City of Morrison, Illinois. The variance allows the City to discharge ammonia nitrogen above the water quality standard into Rock Creek from its wastewater treatment facility. The reason for the variance is the need to repair one of two trickling filters. The variance duration includes time to repair and service the equipment and time to allow the nitrifying bacteria to grow.

The Illinois EPA determined the "environmental impact from the variance should be minimal, no reasonable alternative appears to be available, no public water supplies should be affected, no federal regulations would prohibit granting the request and the City of Morrison would face an arbitrary and unreasonable hardship if the Illinois EPA did not grant the requested variance" (IGNN, 2006).

2.4.2 Example Phosphorus Variance

An example of a phosphorus variance is the variance granted the City of McKinley, Minnesota. The variance allows the city to discharge total phosphorus above the water quality standard into an unnamed creek from its wastewater treatment plant. The city requested the variance for economic reasons and a shrinking population base.

If the city were required to meet a 1 mg/L phosphorus limit, it would need to add phosphorus removal equipment to their current facility. This would require raising the assessed fees to individual users. Treatment alternatives evaluated included regionalization, a mechanical plant, stabilization ponds with alum, and a sand filter. The calculations suggested that treatment could be achieved for an increase of less than $5 per month.

The State's assessment considered the economic conditions of the city, its small population base, and its limited financial resources which linked this with the substantial and widespread economic and social impact factor. The Minnesota Pollution Control Agency granted the variance not for undue economic burden, but because they determined that strict conformity with the standard was unreasonable. Although the city's discharge has a phosphorus concentration between 1 and 6.5 mg/L, the State determined that it would not create significant environmental effects. The variance continues to be renewed.

2.5 Narrative Nutrient Standards – Historical

Water quality legislation in the United States began in the 19th century. Early legislation included the Rivers and Harbors Act of 1899, with regulations on the dumping of refuse into waterways. In 1948, key legislation included the Federal Water Pollution Control Act. The states maintained responsibility for water quality. Each state could develop water quality legislation; however, due to the lack of data and supporting science, few standards were adopted and even fewer were enforced.

Significant amendments to the Federal Water Pollution Control Act occurred in the 1970s. Among these changes were the 1972 amendments, commonly referred to as the Clean Water Act. The Act included many new requirements, including setting total maximum daily loads and point source discharge requirements. The Act also reinforced the earlier legislation of states setting water quality standards.

The objective of the Act, which is to restore and maintain the chemical, physical, and biological integrity of the nation's waters, was and still is a huge task. In the following years, states generally found it easiest to focus on point sources and conventional and toxic pollutants for which impairments were observed. Consequently, nutrients became a lower priority. Additionally, there were no pressures to meet the challenge of setting specific numeric nutrient criteria.

As a result, states generally met the minimum requirements of the Clean Water Act by including simple narrative standards addressing nutrients. The following are examples of such narrative standards:

- Surface waters of the state shall be free from excess nutrients that can cause visible slime growths or other aquatic growths impairing designated beneficial uses (IDAPA 58.01.02).
- Except as due to natural conditions, nutrients shall not be allowed in concentrations that cause objectionable algal densities, nuisance aquatic vegetation, abnormal diurnal fluctuations in dissolved oxygen or pH, changes to the composition of aquatic ecosystems, or otherwise render the waters unsuitable for the designated uses (NJAC 7:9B).

A survey of the states and entities authorized to promulgate water quality standards revealed these narratives are still in place. "Every state had narrative criteria that protected the waters from 'objectionable' conditions, which indirectly targets nutrients. Other states had narrative nutrient criteria, which specifically mentioned eutrophication as a problem to be prevented in their narrative nutrient standard" (EPA, 2003).

The states have a long history with setting water quality standards. Over the last 50 years, every state had developed and enacted narrative nutrient standards. The states are likely to maintain these narrative standards as they add new numeric requirements.

2.6 Numeric Nutrient Standards Initiative – Emerging

The states found that the narrative nutrient standards were too broad to be effective and too cumbersome to apply to specific waterbodies. In the meantime, the identification of impaired waters, 303(d) listings, surged past 10,000 nutrient and nutrient-related listings, with 15 states having more than 200 and only one state not having any (EPA, 2003). The EPA recognized the difficulty in using narrative standard to develop TMDLS and NPDES permits, assessing monitoring programs, setting measurable objectives, and evaluating effectiveness. In 1995, the EPA found that most states did not have effective nutrient standards. This was the start of the numeric nutrient standards initiative.

In June 1998, the EPA published the National Strategy for Development of Regional Nutrient Criteria. A key component of the strategy was that the EPA would develop waterbody-type technical guidance documents. The EPA's previous guidance documents were the Quality Criteria for Water, known as the Red and Gold Books. These documents included the criteria for nitrates/nitrites of 10 mg/L (as NO_3-N) for domestic water supply (health) and 0.10 μg/L yellow (elemental) phosphorus for marine or estuarine water. EPA published technical guidance for developing criteria for lakes and reservoirs in May 2000, rivers and streams in June 2000, and estuaries and coastal waters in October 2001 and recommended nutrient criteria for most streams and lakes in January 2001 (EPA, 2003).

In November 2001, EPA issued a memorandum to the states about planning the development and adoption of nutrient criteria into water quality standards. Key elements included the recommendation that the states develop local nutrient criteria based on the technical guidance manual processes. This was the preferred approach, although the states could adopt the EPA 's recommended section 304(a) criteria for nutrients or develop their own scientifically defensible criteria. The EPA expected the states to have a plan by 2004 that outlined their approach to implementing the nutrient criteria.

Table 2-1 shows that some states have finished developing criteria for rivers and streams while others are just starting. States may have also developed criteria for other waterbody types, including lakes, estuaries and wetlands.

Table 2-1. Status of States and Territories Numeric Nutrient Criteria (U.S. EPA, 2007 Status Report).[a]

Rivers and Streams

Stage	Number	States
Has approved criteria for all parameters	5	TN, HI, AS, GU, CN
Has approved criteria for N, P, or Chlorophyll	4	DC, FL, OK*, NV
Engaged in developing criteria for all parameters and waters	6	MA, ME, VT, KY, MI, WI
Collecting data for all parameters or water	34	CT, NH, RI, NJ, NY, PR, DE, MD, PA, VA, AL, FL, GA, MS, NC, SC, IL, IN MN, OH, AR, LA, OK, NM, TX, IA, KS, MO, NE, CO MT, UT, AZ, CA
Just starting criteria process	8	WV, ND, SD, WY, AK, ID OR, WA

*OK: scenic rivers only,
Updated May 14, 2007

[a] The EPA 2008 status report is at http://www.epa.gov/waterscience/criteria/nutrient/files/report1998-2008.pdf

2.7 EPA Ecoregion Reference Criteria

The ecoregion criteria concept was included in the National Strategy for the Development of Regional Nutrient Criteria from the beginning, citing the work of James Omernik of the EPA Corvalis, OR laboratory. The concept of ecological regions, or ecoregions, is the grouping of areas of similar climate, hydrology, geology, physiography, soils, land use, vegetation, and wildlife. There are four levels of ecoregions, with Level I being the coarsest and Level IV the most detailed. Fourteen ecoregions in the continental United States are Level I, while 104 are Level III.

The EPA has recommended criteria for total phosphorus and total nitrogen for Level III ecoregions that have been aggregated into 14 nutrient ecoregions for rivers and streams, lakes and reservoirs, and wetlands. "The nutrient criteria presented by the EPA for each ecoregion are generally based on the 25th percentile value of all data from the respective ecoregion. The 25th percentile value corresponds to the concentration at which 25% of the measured values are below and 75%t of the measured values are above" (EPA, 2000a). A summary of the rivers and streams criteria are shown in Table 2-2.

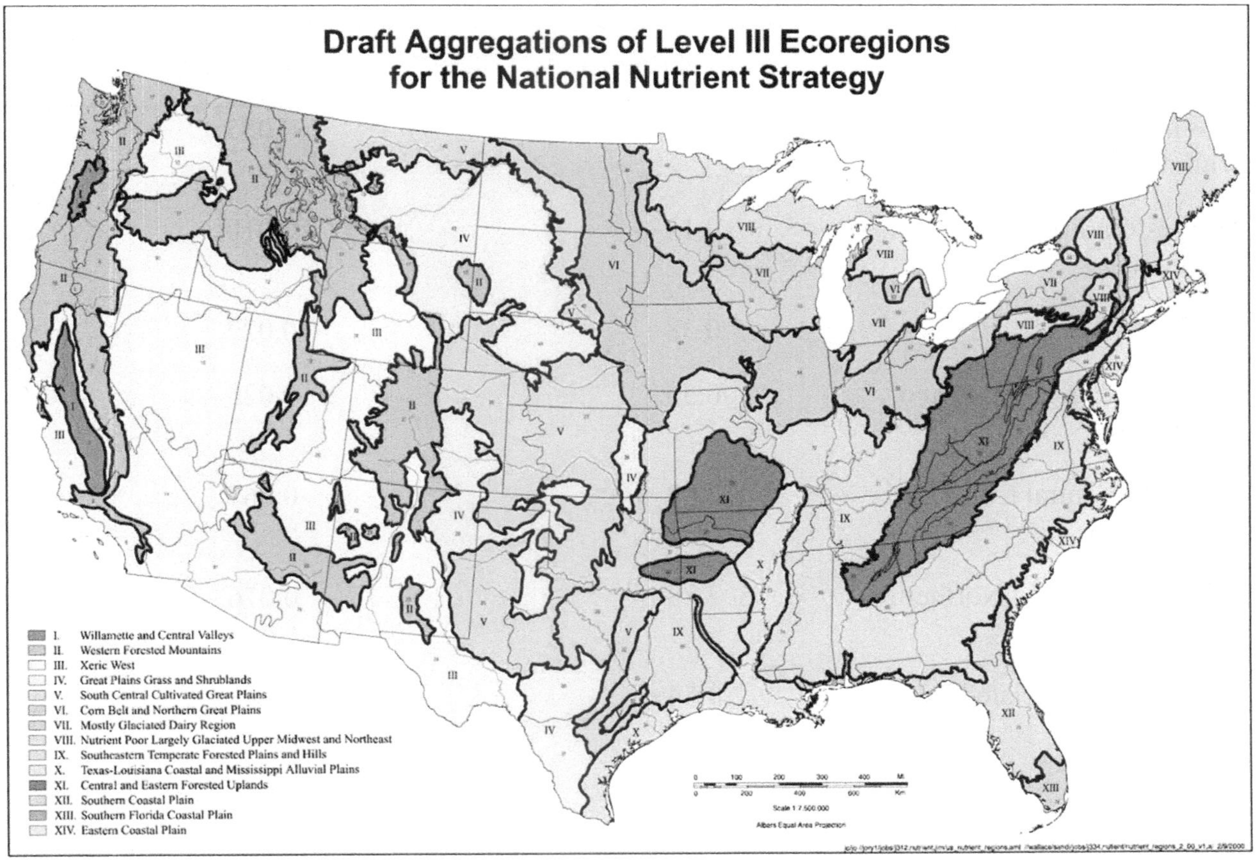

Figure 2-1. U.S. EPA Aggregate Level III Ecoregions.

Table 2-2. Summary of Ecoregion Criteria for Rivers and Streams.[a]

Ecoregion	TN (mg/L)	TP (mg/L)
I: Willamette and Central Valley	0.66	0.055
II: Western Forested Mountains	0.12	0.010
III: Xeric West	0.38	0.022
IV: Great Plains Grass and Shrublands	0.56	0.023
V: South Central Cultivated Great Plains	0.88	0.067
VI: Corn Belt and Northern Great Plains	2.18	0.076
VII: Mostly Glaciated Dairy Region	0.54	0.033
VIII: Nutrient-Poor, Largely Glaciated Upper Midwest and Northeast	0.38	0.010
IX: Southeastern Temperature Forested Plains and Hills	0.69	0.037
X: Texas-Louisiana Coastal and Mississippi Alluvial Plains	0.57	0.060
XI: The Central and Eastern Forested Uplands	0.31	0.010
XII: Southeastern Coastal Plain	0.90	0.040
XIII: Southern Florida Coastal Plain	1.14	0.015
XIV: Eastern Coastal Plain	0.71	0.031

[a] Sources: http://www.epa.gov/waterscience/criteria/nutrient/ecoregions/index.html with corrections from the federal register corrections http://www.epa.gov/fedrgstr/EPA-WATER/2003/January/Day-06/w176.htm

In addition to the nutrient criteria, the EPA developed chlorophyll and water clarity – as either turbidity or Secchi depth – criteria for the Level I ecoregions. These causal and response criteria were meant to be starting points for the States to develop local numeric criteria to address local eutrophication problems. These local criteria should then be protective of the beneficial uses assigned to the waterbody.

2.8 Nutrient Standards Based on Invertebrates and Related Water-Quality Parameters

Nutrient standards may also be based either indirectly on invertebrates and other related water-quality parameters such as dissolved oxygen and their impact on the aquatic ecosystem. Invertebrates for example, are frequently part of a bio-assessment and the determination of impairment. Maintaining the diversity of invertebrates is part of the Clean Water Act goal of protecting ecological integrity, fish, shellfish, and wildlife. All stressor-response, modeling, and distribution based reference approaches to nutrient standards are recent evolutions which required further development. While water quality and invertebrates have long standing linkages, their direct connection to nutrient standards has received a great deal of attention recently.

The terms "invertebrates," "aquatic invertebrates," and "macroinvertebrates" traditionally refer to spineless creatures, such as insects that inhabit a river channel, pond, lake, wetland or ocean and are visible to the eye. Flow, water quality, food (prey/predators), and habitat influence invertebrate abundance and diversity. This makes invertebrates an integrated ecosystem health indicator. The population of certain species of invertebrates correlates with the water quality. For example, stoneflies are often associated with clean water, while worms and midges are associated with polluted waters.

A cause and effect relationship provides a basis for establishing a nutrient standard. For example, the discharge of too high a nutrient loading (the cause) results in too much algae (the effect) and impairs the water for fishing, swimming, drinking, etc. While water temperature and sunlight also affect the growth of algae, nutrients are a fundamental building block of algae growth and there is a direct cause and effect relationship. Nutrients standards may also be set based on dissolved oxygen concentrations. Since the growth and decay of algae affect dissolved oxygen concentrations, the cause and effect relationship is only one-step further removed.

Nutrient concentrations can influence the entire aquatic ecology of a region. Nutrients can cause low dissolved oxygen concentrations, which impact fish, algae blooms that alter habitat, algae speciation shifts, biodiversity changes, dissolved oxygen and pH swings, and variations in the conditions for macroinvertebrates, fish, and mussels. However, characterizing the response may be challenging due to uncertainty and the influence of other environmental factors, such as temperature, the amount and intensity of sunlight, the depth of the water body, water movement, and the nitrogen to phosphorus ratio. Appropriate statistical analysis can be used to reduce the level of uncertainty. If the uncertainty is unacceptably high, then mechanistic models and reference conditions can be utilized.

Additional challenges to appropriately correlating nutrients to ecosystem health include selecting standards that balance a variety of objectives. "The ultimate goal of nutrient criteria development is protection of these designated uses. However, the correlation of nutrient loading with use impairment is not always clear. In the case of protection of aquatic life, a certain amount of nutrient loading is desirable. Without it, there is insufficient phytoplankton and macrophyte growth to support a robust food chain. On the other hand, excessive nutrient loading has been associated with toxins produced by cyanobacteria, as well as some fish kills that resulted from loss of oxygen due to plant decay. The range of desirable nutrient concentrations for striking the right balance between these two undesirable situations will vary, depending on lake morphology and the types of aquatic species that are present, among other factors." (Missouri, 2005).

The EPA published guidance on linking nutrient availability to algal response, and developing predictive relationships, as part of the Nutrient Criteria Technical Guidance Manual (U.S. EPA, 2000b). They recognized the same challenges, especially for relating nutrients to invertebrates and higher life forms. "Effects of nutrients on algal biomass and effects of algae on the biotic integrity of macroinvertebrates and fish should be characterized to aid in developing nutrient criteria that will protect designated uses related to aquatic life." However, the EPA 's recommendation was not to try to make direct connections. "It is recommended that relations between biotic integrity of algal assemblages and nutrients be defined and then related to biotic integrity of macroinvertebrate and fish assemblages in a stepwise, mechanistic fashion."

States are already struggling with the development of numeric nutrient standards and incorporating invertebrates may pose an additional challenge. However, EPA provides professional statistical analysis for those States who have the invertebrate data that can be used to identify relationships with nutrients to strengthen the scientific defensibility of the nutrient endpoints proposed in state standards. States which have strong biocriteria programs are in the best position to utilize nutrient invertebrate relationships.

Minnesota and Wisconsin have been examining the relationships among nutrients and biology, searching for potential indicators. Indiana has examined the cause and effect relationships but has not found any strong and consistent correlations. West Virginia also found weak relationships between nutrient effects and macroinvertebrates. Michigan and Ohio have found some correlations but decided they were not sufficient to create standards and instead developed standards based on distributions. Additionally, Ohio has developed an index of biotic integrity. New York, New Jersey, and Pennsylvania have also discovered some correlations. Finally, Florida has developed relationships between algal species indicators and nutrient conditions. "Species response measures were found to best correlate with a multimetric index describing nutrient concentrations, incorporating values for five water chemistry parameters (NH3, NOx, TN, PO4, and TP) and three measures of algal biomass response (periphyton and plankton chlorophyll a and assays of algal growth potential)." Again, although there are correlations, Florida is not using them to set nutrient standards.

Nutrient to invertebrate relationships are complex. The cause and effect relationships may be masked, hidden or altered by a long list of factors. This complex interconnectivity among nutrients, algae, oxygen, biological communities and physical habitat confounds the analyses. This has lead to some major disagreements about the use of macroinvertebrate indices for nutrient target setting.

While the EPA has promoted this methodology of linking algal growth and nutrients, there are still issues to resolve, including scientific peer review and acceptance (Hall, 2009). "Existing guidance was premised on demonstrating that nutrients are causing excessive plant growth and TMDLs are only developed where site-specific information confirms that a problem exists. This new approach, approved by EPA Headquarters, presumes nutrients directly impair invertebrate communities." (Lee, 2009).

Setting nutrient standards based on invertebrates requires further research and critical examination. It is currently unclear if scientifically defensible relationships can be found and how the EPA will proceed with the formulation of guidelines. EPA is developing guidance on empirical approaches for stressor-response analyses using a variety of responses including invertebrates and this is undergoing peer review. Use of similar analytical approaches for clean sediment have been peer reviewed and published.

2.9 EPA Science Advisory Board's Review of EPA Draft Guidance

The EPA Office of Water published draft guidance on using empirically derived stressor-response relationships as the basis for developing numeric nutrient endpoints for water quality standards in response to the interest of many states. The draft document titled "Technical Guidance on Empirical Approaches for Numerical Nutrient Criteria Development" (EPA, 2009a) was intended to supplement EPA 's published technical guidance for developing numeric nutrient water quality standards, which focuses on the use of reference conditions for deriving nutrient criteria.

In August 2009, EPA appointed a Science Advisory Board (SAB) to review the draft Empirical Approaches for Nutrient Criteria Derivation. The SAB Ecological Processes and Effects Committee, comprised of eleven members, mostly academics and a few consultants, performed the review and issued draft results in January 2010, followed by a final report in April 2010 (EPA, 2010d).

The EPA Office of Water believes numeric nutrient water quality standards are an important goal and has set a high priority for state adoption of numeric nutrient criteria. EPA states that this is an important goal for the following reasons: supports the development of nutrient related TMDLs, provide targets for nutrient trading programs, make it easier to write NPDES permits, evaluate the success of nutrient runoff reduction programs, and measure water quality progress.

The SAB Committee reported that the guidance lacks an explicit and direct explanation of how it links to, and supports the goals of the Clean Water Act. While the SAB Committee agreed with the importance of EPA's efforts to control and reduce nutrient pollution; the SAB Committee found the guidance document was not ready for use, being incomplete and confusing to implement. Additionally, it was not clear to the SAB Committee how the guidance document integrates with EPA's existing nutrient criteria technical guidance manuals and documents.

The intended audience for the EPA guidance is state and tribal water quality scientists and resource managers for their use in developing numeric nutrient criteria. The SAB Committee found that the guidance was too technical for most audience members and would require significant training and additional examples are needed to be helpful. There was concern expressed about the risk that the guidance could not be easily followed, resulting in misuse and misapplication. The guidance was found to be imprecise in leading managers to the selection of the most appropriate and defensible criteria. The SAB Committee found that the guidance did not address or solve the factors that have limited progress toward developing nutrient criteria, such as potentially limited availability of data, lack of available technical expertise, insufficient resources, and expense. The guidance may provide little help towards accelerating the adoption of numeric nutrient criteria.

The SAB Committee agreed that the stressor-response approach is a means for developing numeric nutrient criteria if used with other methods and applied appropriately. The

SAB Committee warned against using statistical methods without an understanding of the technique. The SAB Committee emphasized that statistical associations may not be biologically relevant and the methods proposed do not prove, or establish, cause and effect relationships. The findings summarized this absence of a direct causative relationship between stressor and response as one of the most important review comments.

The SAB Committee found that the EPA guidance did not sufficiently address uncertainty in the data and analysis methods. There may be significant variation in using empirical stressor-response approaches to establish criteria. The SAB Committee found that the EPA guidance is not clear on how to select numeric criteria when various analyses provide a range of results. The SAB Committee stated that there was insufficient detail regarding how to examine the accuracy, reliability, and validity of the results of using the methods. There was concern that if the proposed methods yielded inaccurate results, it could lead to inappropriate, or ineffectual solutions, and the selection of criteria could result in severe environmental, social and economic consequences.

The SAB Committee found that the EPA guidance fails to follow the principle of a watershed approach to managing water quality issues. The methods in the guidance do not address the problem of excess nutrient enrichment in downstream waters. Also, there are additional statistical, visualization and other tools that are available and useful for examining nutrients that are not presented in the EPA guidance.

The EPA guidance focused on total nitrogen and total phosphorus. The SAB Committee recommended adding more emphasis on biologically available nutrients. Additionally, they suggested more emphasis on analysis to recognize co-limitation by both nitrogen and phosphorus, which may be common in many systems and regions. The SAB Committee noted that it is important to recognize that single variable stressor-response relationships are rare and there may be systems where nutrient concentrations are not even appropriate stressor variables. The SAB Committee also found that the guidance focuses too heavily on nutrient response relationships driven by autotrophic processes and should include consideration of other variables and processes, including heterotrophic microbes and detritus based systems.

The SAB Committee found that the discussion about interpreting the temporal and spatial aspects of water quality data and establishing relationships was inadequate. This includes establishing linkages between concentrations, loads, and biomass. Attempting to select nutrient concentration criteria based on point-in-time and point-in-space data to influence biomass driven by nutrient supply rates and mass loads may not be possible. Conversely, attempting to select nutrient criteria from load-response models has the same problems.

The EPA guidance does not consider the direct and indirect effects that reduction activities, such as best management practices, may have on stress-response relationship. For example, activities to reduce nutrient loads may also influence the overall ecology of the waterbody and completely change, rather than shift the nutrient dynamics. The example provided was adding stream buffers which filter nutrients but change the riparian vegetation and streambank characteristics, change sediment loading and channel bed materials, may increase shading, and impact the system more than just reducing nutrients. The SAB Committee found that the implications of reduction activities should be included as part of consideration of the overall watershed.

The SAB Committee review included many recommendations to address concerns ranging from specific statistical methods, to the big picture implications of the resulting decisions. The guidance failed to provide the linkage between impairment, designated uses, and management and regulatory targets to meet Clean Water Act objectives. Without revisions, the SAB Committee found that the current draft EPA guidance may not be as instructive to resource managers as intended for accelerating the development of numeric nutrient criteria.

2.10 State Numeric Nutrient Standards

A number of states are in the process of developing a basis for numeric nutrient standards, notably Montana, Colorado, and Wisconsin. EPA has proposed numeric nutrient standards for Florida rivers and streams, as discussed later in this chapter under the heading Nutrient Related Legal Issues.

2.10.1 Montana

The State of Montana has been developing numeric nutrient criteria to control excessive nutrient (nitrogen and phosphorus) pollution in Montana's streams, rivers, and lakes since the early 2000's. Based on the studies completed by the State, the Montana Department of Environmental Quality is initiating a rule-making phase for state adoption of numeric nutrient standards (MDEQ, 2010a).

In EPA's State Adoption of Numeric Nutrients Standards, Montana is identified as having existing numeric water quality standards for nutrients only for selected rivers and streams (MDEQ, 2008a). In Montana, Numeric Water Quality Standards Circular DEQ-7 is incorporated by reference into the Administrative Rules of the State of Montana (ARM) 17.30.619 (MDEQ, 2008b, MDEQ, 2010b). No values are cited for phosphorus and nitrogen only the footnote "A plant nutrient, excessive amounts of which may cause violations of Administrative Rules of Montana (ARM) 17.30.637 (1)(e)" (MDEQ, 2008b).

Included within ARM 17.30.631 are numeric algal and biomass and nutrient standards. Numeric criteria are provided for two reaches of the Clark Fork River in western Montana. In the mainstem Clark Fork River from below the Warm Springs Creek confluence to the confluence with the Blackfoot River the numeric water quality standard is 300 ug/L for total nitrogen and 20 ug/L for total phosphorus applicable from June 21 to September 21. In the Clark Fork River from the confluence with the Blackfoot River to the confluence with the Flathead River the numeric water quality standard is 300 ug/L for total nitrogen and 39 ug/L for total phosphorus applicable from June 21 to September 21. These values were a result of the Clark Fork River Voluntary Nutrient Reduction Program (VNRP), a TMDL initiated in 1994 and completed in 1998 as an agreement among major parties in the Montana portion of the watershed to significantly reduce nutrient pollution along a 200-mile stretch of the Clark Fork River (TSWQC, 2010). The VNRP included the formation of a Nutrient Target Committee by the Tri-State Water Quality Council, aimed at achieving consensus on in-stream nutrient targets for the Clark Fork River and to develop a basin-wide nutrient source reduction program to meet those targets.

Montana's approach to the development of numeric nutrient standards has included review of reference stream criteria as well as site specific investigations such as the Clark Fork River studies discussed above, and other studies focused on cause and effect relationships. The existing narrative and numeric criteria address only the effect variables and do not address the root cause of the effects. "*Numeric nutrient criteria will improve upon the existing standards because they address the causes of eutrophication directly*" (MDEQ, 2008a).

The MDEQ and University of Montana conducted a public survey of perceptions of stream health and bottom algae (UMT, 2007). Eight photographs of streams with various bottom algae conditions were used in the survey. Respondents to the survey were asked whether the conditions in each photograph were desirable or undesirable. The conclusions reached from the survey results were that at chlorophyll levels of 200 mg/m^2 and greater, the condition is undesirable and that nutrient standards should be set at the concentrations associated with stream bottom chlorophyll levels of approximately 150 mg/m^2.

Other studies have been completed by MDEQ, including the Scientific and Technical Basis of the Numeric Nutrient Criteria for Montana's Wadeable Streams and Rivers which was peer reviewed (MDEQ, 2008a). The reviewers concluded that the study was "*scientifically rigorous, well documented, thoughtful, and thorough.*" As a result, the reviewers expressed confidence that the procedures presented in the document would result in defensible nutrient criteria. The reviewers essentially endorsed Montana's approach and felt it offered a sound scientific basis for developing nutrient criteria for wadeable streams" (TetraTech, 2008). One of the first studies completed was the Wadeable Streams of Montana's Hi-line Region: An Analysis of Their Nature and Condition, with an Emphasis on Factors Affecting Aquatic Plant Communities and Recommendations to Prevent Nuisance Algae Conditions (MDEQ, 2004). "*The main objective of this study was to define nuisance levels for benthic and other algae in wadeable streams of Montana's Hi-line region, and to identify those factors that control algal and other aquatic plant (e.g., macrophyte) biomass*" (MDEQ, 2004). The findings recommend a total nitrogen concentration of less than 1.044 mg/L and a total phosphorus concentration of less than 153 ug/L. The findings also recommend low nitrate/nitrite and SRP concentrations of 2.3 ug/L and 8 ug/L, respectively.

Montana's recommended wadeable stream nutrient criteria are summarized in Table 2-3.

Table 2-3. Montana Department of Environmental Quality Recommended Numeric Nutrient and Benthic Algae Criteria for Wadeable Streams and Rivers in Ecoregions of Montana.

		Nutrient Criteria			
Level III Ecoregion	Period When Criteria Apply	Total P (mg/L)	Total N (mg/L)	$NO_{2\text{-}3}$ (mg/L)	Benthic Algae Criteria
Northern Rockies	July 1 – Sept. 30	0.012	0.233	0.081	150 mg Chl *a*/m^2 (36 g AFDW/m^2)
Canadian Rockies	July 1 – Sept. 30	0.006	0.209	0.020	150 mg Chl *a*/m^2 (36 g AFDW/m^2)
Middle Rockies	July 1 – Sept. 30	0.048	0.320	0.100	150 mg Chl *a*/m^2 (36 g AFDW/m^2)
Idaho Batholith	July 1 – Sept. 30	0.011	0.130	0.049	150 mg Chl *a*/m^2 (36 g AFDW/m^2)
Northwestern Glaciated Plains	June 16 – Sept. 30	0.123	1.311	0.020	n/a
Northwestern Great Plains, Wyoming Basin	July 1 – Sept. 30	0.124	1.358	0.076	n/a

Large river nutrient criteria have also been investigated in Montana. West of the continental divide, Clark Fork River studies conducted in the 1990's explored benthic algae density and threshold nutrient concentrations and resulted in the adoption of site specific nutrient standards. East of the continental divide, MDEQ is conducting a mechanistic modeling study of the Yellowstone River in the reach downstream of Billing, Montana and extending towards Miles City and Glendive, Montana. The focus of the study is to investigate appropriate numeric nutrient standards for a large river that has a variety of influences that make it unique (east of the continental divide, extending into the plains, changes in channel morphology, increased turbidity from land use and tributaries) (USGS, 2001).

The MDEQ modeling study of the Yellowstone River downstream of Billings has combined a field monitoring effort with QUAL2K modeling using a new version of the model that includes simulation of periphyton. The model has been calibrated and MDEQ plans to use it as a tool to establish target in-stream target concentrations for nitrogen and phosphorus. At this time, consideration has not been given to implications for effluent discharge permit limits that might be associated with in-stream nutrient targets on the Yellowstone River, however 15Q10 river flows conditions have been used in water quality model simulations to associated a low flow critical water quality condition with the expected algal response time in Montana watersheds.

2.10.2 Colorado

In 2002, the Colorado Department of Public Health and Environment (CDPHE) initiated a nutrient criteria development plan in response to EPA's recommendation to adopt nutrient criteria (CDPHE, 2002). The State of Colorado initiated preparation of numeric nutrient criteria with a decision not to adopt EPA ecoregion reference criteria in the interest of developing more site-specific standards. In addition, Colorado sought a more direct linkage to the causes and responses of the waterbody to increased nutrient loads. While some nutrients are necessary to support aquatic life and fisheries, Colorado's focus for developing numeric nutrient standards is to prevent hyper-enrichment that leads to eutrophication.

The method Colorado used to develop potential nutrient standards differed from the statistical approach EPA followed to develop ecoregion reference criteria. Not only were lakes, reservoirs, rivers and streams evaluated separately, but Colorado also examined inter-basin transfers and water supply reservoirs separately. Colorado used algae/chlorophyll and aquatic life use support as key indicators (CDPHE, 2007). Colorado has developed some site specific nutrient criteria (EPA, 2006).

Colorado appears to have made significant progress since in developing nutrient criteria and is working towards integrating those criteria into their water quality standards. The Colorado Water Quality Forum (2010) provides an overview of the process to develop standards. The Colorado Water Quality Control Commission has decided to delay consideration of numeric nutrient criteria for rivers and stream, lakes and reservoirs, and direct-use water supply reservoirs until June 2011 (CDPHE, 2010a).

The CDPHE approach is based on defining nutrient criteria with a stressor-response relationship linking nutrient concentrations with Colorado's macroinvertebrate multi-metric indices (MMIs). In February 2010, CDPHE proposed nutrient criteria based on aquatic life use protection. Table 2-4 summarizes the initial values developed for rivers and streams with total phosphorus of 90 µg/L and total nitrogen of 824 µg/L for cold water biota streams, and total phosphorus of 135 µg/L and total nitrogen of 1,316 µg/L for warm water biota streams (CDPHE, 2010b). Colorado's acute and chronic standards include the concept of protecting 95% of the genera from acute or chronic effects of the parameter. This concept is included in the CDPHE proposal for numeric nutrient standards is to select criteria that allow a 5% decrease in biological conditions based on an Observable Biological Potential ("OBP") that describes the decline in biological condition as a function of increasing nutrient concentrations. The nutrient criterion is to be determined by locating the concentration at which the OBP is expected to be 5% below the reference condition, with the anchor point for the allowable decline set at the 85th percentile concentration of the set of reference sites. The CDPHE intends to implement these standards in the same manner that chronic total metals standards are implemented, with a one in three year allowable exceedance frequency with permits based on a 30-day average.

Table 2-4. Initial Proposal for Colorado Nutrient Criteria for Rivers and Streams. (CDPHE, 2010b)

	Cold Water Biota		Warm Water Biota	
	TP (µg/L)	TN (µg/L)	TP (µg/L)	TN (µg/L)
Possible Criteria [a] (based on fit of Observable Biological Potential (OBP) line)	90	824	135	1,316
Upper and Lower confidence limits [b]	82-129	776-988	125-184	1,251-1,538

[a] The "possible criteria" represents the Division's best estimate of the criteria at this point in the process
[b] These confidence limits are provided to illustrate the confidence bands on the OBP line

2.10.3 Wisconsin

The State of Wisconsin is in the process of developing water quality criteria for phosphorus. The development of these phosphorus criteria are identified as an item of "Group A: Revisions/Development Currently in Progress" for the 2008-2011 Triennial Standards Review Cycle (DNR, 2010a). While the State does not have nitrogen on the priority list, nitrogen is also expected to require examination. Wisconsin may implement nutrient standards in the next few years through either the State rulemaking process, or potentially as promulgated by EPA as part of a Gulf of Mexico nutrient reduction plan.

The State of Wisconsin Department of Natural Resources (DNR) and the United States Geological Survey (USGS) have completed numerous water quality studies in Wisconsin. USGS studies include investigation of the relationships between nutrient concentrations and the biotic integrity of nonwadeable rivers and wadeable streams (USGS, 2006, USGS, 2008). This extensive knowledge of water quality, along with the DNR's water quality monitoring database, the Surface Water Integrated Monitoring System (SWIMS), have been cited as the basis for water quality impairment. The Wisconsin 2010 303(d) list includes 1,216 individual 303(d) listings for 523 waterbodies (DNR, 2010b).

The DNR formed a technical advisory committee to assist with development of phosphorus criteria and to review draft rules for nutrient standards. The proposed phosphorus criteria developed for streams is 0.075 mg/L and for large rivers is 0.100 mg/L (UMRCC, 2008). The proposed phosphorus criteria developed for lakes and reservoirs varies from 0.015 mg/L to 0.40 mg/L depending upon stratification and drainage characteristics (DNR, 2009). The criteria are based on studies completed on Wisconsin waterbodies, along with scientific concepts for river and lake water quality (Clean Water, 2008). The proposed draft rules may be incorporated in NR 102 water quality standards for Wisconsin surface waters and NR 106 procedures for calculating water quality based effluent limitations for toxic and organoleptic substances discharged to surface waters (DNR, 2008). In mid-2009 the draft rules were under review by the technical advisory committee (CSWEA, 2009a, CSWEA 2009b).

In late 2009, a coalition of environmental groups announced their intent to sue EPA to promulgate numeric nutrient criteria for phosphorus and nitrogen for the State of Wisconsin (MEA, 2009a, MEA, 2009b). The group stated the need to accelerate the process and enact standards. The group also stated that the DNR has developed the science needed for sound phosphorus standards. The notice of intent to sue includes, "*DNR has yet to propose that its governing board, the Natural Resources Board, amend the Wisconsin Administrative Code to include numeric criteria for phosphorus. DNR does not expect to begin promulgation of numeric nitrogen water quality criteria until at least 2012. In the meantime DNR refuses to derive water quality based effluent limits in NPDES permits to implement its narrative standard as applied to nitrogen and phosphorus*" (MEA, 2009b).

2.11 Nutrient Innovation Task Group

In 2008, an ad hoc State-EPA Nutrient Innovation Task Group (NITG) was formed to review both existing and innovative approaches to nutrient (nitrogen and phosphorus) management. The charge was to evaluate the national nutrient issue and frame innovative solutions for greater results. Goals of the group were "to identify and frame key nutrient issues, questions, and options on how to improve and accelerate nutrient pollution prevention and reduction at the state and national level" (EPA, 2009b). NITG is a collaborate effort of EPA Office of Water and state representatives. Participants included one representative each from nine states (CT, DE, IL, KS, OK, OH, UT, VA, WI), with one each from three associations (ASIWPCA, ORSANCO, ASDWA), one each from nine of the 10 EPA regions, and five from the EPA Office of Water.

To find innovative approaches to deal with nutrients, the group was subdivided into the following subgroups and task activities:

A – Characterization and Scope of Nutrients Problem
B – Review and Analysis of Existing and Alternative Tools
C – Existing and Future Cost of Control, Status Quo, or Restoration
D – How to Better Communicate the Issues and Implications of Excess Nutrients
E – Evaluate New Methods of Accountability including Control of Diffuse Pollution Sources

The report concludes that efforts to slow and reverse the degradation of drinking water supply and natural resources has been inadequate and such degradation will even accelerate with continued development and population growth over the next 40 years: "*States and EPA recognize that eutrophication and nutrient overloading are significant environmental problems, not just for aquatic resources but also from a drinking water standpoint. In the past, we have been successful in some areas, but not in others. We agree to meet to develop a strategy to change the way we act to improve ways to reduce or eliminate nutrient releases*" (U.S. EPA,

2009c). Nutrients threaten every aspect of our water including public health endangerment, drinking water impairments, water quality impairments, and socioeconomic considerations.

NITG relied on the extensive documentation and studies on the impacts of excessive nitrogen and phosphorus on the nation's waters. Nitrogen and phosphorus pollution is significant across the country, with 49 states having nutrient impairment listings. The group synthesized and examined this existing information to propose new, innovative tools to improve control of nutrient pollution sources.

The scope of nutrient pollution is large, significantly impacting drinking water supplies, aquatic life and recreational water quality. A portrayal of these follows (EPA, 2009c):

There are over 52,000 community water systems serving more than 290 million people.

Protecting the water quality of these sources is in our best interest rather than dealing with the cost and complexity of treatment.

Nutrient pollution is directly linked to 20% of impaired river and stream miles, 22% of impaired lake acres, and 8%t of impaired bay and estuarine square miles.

Including impairment indirectly to nutrients these increase to 31% of impaired river and stream miles, 30% of impaired lake acres, and 50% of impaired bay and estuarine square miles.

Nutrient pollution from a variety of sources was examined by the NITG including urban and suburban stormwater runoff, municipal wastewater discharges, air deposition, and agricultural livestock activities and row-crop runoff (EPA, 2009c). Much of the U.S. population is consolidated in urban areas, with over 80% of people living on about 66 million acres of land. Urban stormwater is challenging due to the higher annual volume of runoff and the additional and variable concentrations of nonpoint source pollutant concentrations. Municipal wastewater facilities treat about 34 billion gallons of wastewater per day and some 18 million tons of solid waste a year. Treating nutrients to the technology based limits of 3 mg/L for nitrogen and 0.1 mg/L for phosphorus was estimated to cost about $44 billion each or $54 billion simultaneously. Another aspect of wastewater is about 20% of U.S. homes use septic systems which are significant contributors of nutrients especially as they age. Livestock is an even larger challenge with some 96 million head of cattle, 68 million head of swine, 9 billion broilers and 446 laying hens generating more than 1 billion tons of manure a year. This manure results in over 8 million pounds per day of nitrogen and 3 million pounds per day of phosphorus. Additionally significant aspects of agriculture manure management remain generally unregulated. Agricultural row-crops are produced on over 300 million acres across the country.

The NTIG "*was unanimous in its assessment that existing CWA tools have not been fully implemented to reduce nutrients*" especially considering the scope of the problem. The group examined nineteen existing tools grouped into four categories, as follows:

- Tools for Point Sources and Non-Point Sources
- Tools for Point Sources
- Tools for Non-Point Sources
- Tools for Other

The NITG found that the distribution of overall usefulness was evenly split, with seven of the tools rated high (mostly for point sources), five as moderate, and seven as low (mostly for nonpoint sources). The most commonly used tools included NPDES permitting (municipal wastewater treatment and urban stormwater), TMDLs, CAFO regulations, land application permits, the Farm Bill, water quality trading, and water quality regulations (water quality standards, 305 assessments, 303(d) listings, and section 319 grants).

The NITG then identified over 35 tools and authorities, including new, partially used and underutilized tools, which could be used to address nutrient pollution. The tools were grouped into two categories: incentive-based and regulatory. The NTIG provided bullet examples of applying the innovative tools to the five sources of nutrients (urban stormwater, POTWs, air deposition, agricultural livestock, and agricultural row-crops). The top five tools are summarized in Table 2-5.

Table 2-5. Five Highest Ranked Innovative Tools by Nutrient Innovation Task Group. (EPA, 2009b)

Type	Tools	Scale of Implementation		Point Source	Non-Point Source
		National	State		
Regulatory	Nonpoint source regulation: Seek legislative change(s) to authorize regulation of nonpoint source pollution to require nonpoint sources to achieve water quality targets and/or technology-based performance standards.	Yes	Yes		Yes
Regulatory	Established technology treatment requirements for nutrients and thereby establish technology based limits for NPDES point sources that discharge nutrients to waters-update secondary treatment requirements.	Yes	Yes	Yes	
Source Reduction	Detergent phosphate bans		Yes	Yes	
Regulatory	Federally promulgate numeric nutrient water quality criteria/standards	Yes	Yes	Yes	
Incentive-based	Green labeling	Yes	Yes	Yes	Yes

However, the NITG recognized that all of the tools, and especially incentive-based tools, have long-term challenges to success including: sustained funding, follow-up accountability and documentation, voluntary programs that allow opting out, and the inequity between sources with some sources being relied upon to do a majority of the reduction. The NITG noted: "*It is fair to point out that the recommendation to seek to regulate nonpoint sources with a similar rigor to that of point sources was judged to be the most effective tool in reducing nutrient loadings to our nation's water since it is broadly recognized that nonpoint sources contributed the bulk of the nutrient loadings to waters and those loading have been the most difficult to control and reduce*" (EPA, 2009c). This has resulted in the "*growing resistance of heavily regulated point sources to accept major increases in required loading reductions when unregulated nonpoint sources that might be contributing substantial nutrient pollution to the same watershed are not held accountable for achieving comparable load reductions*" (EPA, 2009c).

NTIG believes that addressing the nutrient pollution problem will require national leadership. A national multi-sector framework is necessary for accountability from upstream to downstream in watersheds, cross-state regulations, point and non-point sources, and environmental protection legislation. The NITG found that a critical component is having effective state programs with similar economic impacts (from a national to local scale), such that adjoining states do not benefit by having less innovative and effective programs for controlling nutrient pollution. If the response to the urgent call for action is not heeded, then population growth, urbanization, energy needs especially biofuels, and the impacts of climate change will likely result in continue degradation from nutrient pollution and increased litigation.

2.12 Inspector General Report

On August 26, 2009, the EPA Office of Inspector General issued a report critical of EPA's efforts to manage nutrients nationally and in particular cites the lack of progress in State adoption of numeric nutrient standards. The Inspector General made the following recommendations to Pete Silva, Assistant Administrator of Water:

1. Select significant waters of national value which need numeric nutrient standards to meet the Clean Water Act.
2. Set numeric nutrient water quality standards for those waters (Mississippi River and Gulf of Mexico are highlighted).
3. Establish EPA and State accountability for meeting milestones for adoption of numeric nutrient standards in the rest of the nation by requiring States to develop milestones and by EPA review and approval.
4. Establish metrics to gauge progress by states.
5. Ensure that EPA regions annually validate the Water Quality Standards Action Tracking Application (WATA).

The Inspector General report notes that EPA disagrees with recommendations No. 1 and No. 2 but concluded that EPA's past and current strategy has not been effective and that developing another strategic approach would not be responsive to the Inspector General recommendations.

The Inspector General published findings on the review of the EPA numeric nutrient standards program including the following:

> *"In the 11 years since EPA issued its strategy, half the States still had no numeric nutrient standards. States have not been motivated to create these standards because implementing them is costly and often unpopular with various constituencies. EPA has not held the States accountable to committed milestones. The current approach does not assure that States will develop standards that provide adequate protection for downstream waters. Until recently, EPA has not used its Clean Water Act authority to promulgate water quality standards for States."*

"EPA cannot rely on the States alone to ensure that numeric nutrient standards are established. EPA should prioritize States/waters significantly impacted by excess nutrients and determine if it should set the standards."

In particular, the Inspector General report concludes that the EPA program is not making adequate progress:

"EPA's current approach is not working. EPA has relied on the States to develop standards on their own without any meaningful monitoring or control. EPA did not establish priorities, enforceable milestones, or adequate measures to assess progress. States have made minimal progress in developing standards and have not yet considered the impact of their waters on downstream waters. EPA has neither held the States accountable nor used its CWA authorities to promulgate standards. Consequently, EPA is not assured that the States will set numeric nutrient standards or that the standards would provide adequate protection under the CWA for downstream waters."

The Inspector General calls upon EPA to develop a realistic approach with specific emphasis on key waterbodies and singles out the Gulf of Mexico:

"Given the lack of progress and the challenges involved, EPA needs to develop a realistic approach to meet the intent of the CWA that includes priorities and milestones for action. We believe that EPA should prioritize its efforts by addressing waters of national value (e.g., the Gulf of Mexico) requiring a coordinated effort with several States. Using its CWA authority, EPA should determine if numeric nutrient water quality standards are necessary for those waters and apply its recommended criteria. That would allow EPA a baseline to work with the upstream states to develop reasonable standards and milestones."

"In 2008, 10 years after EPA issued its national strategy, the hypoxic zone in the Gulf of Mexico had become the second largest on record and the second largest dead zone in the world."

"Nutrient pollution is widespread and impacts virtually every State. As required by the Section 303(d) of the CWA, States continue to report over 14,000 impairments for nutrient and nutrient-related pollution on their impaired waters lists."

Appendix B of the Inspector General report highlights the Top 10 states contributing nutrients to the Gulf of Mexico (Illinois, Iowa, Indiana, Missouri, Arkansas, Kentucky, Tennessee, Ohio, Mississippi, Nebraska (Nitrogen), Oklahoma (Phosphorus)). The Inspector General report notes that none of these states had considered their impact on the Gulf in developing water quality standards. It is also noted that rather than relying on States to set standards, EPA could promulgate standards for waters of national value, such as the Mississippi River and Gulf of Mexico.

EPA responded to a draft of the Inspector General report in July of 2009 (Appendix C of Inspector General report) and indicated the following:

"When envisioning this approach, we recognize the strategic importance of addressing waters such as the Gulf of Mexico, the Mississippi River Basin and the Chesapeake Bay. We propose that we could develop this strategic approach in 2010."

2.13 Nutrient Related Legal Issues

Third party environmental groups have played an important role in influencing State and federal requirements for nutrient control. The NRDC petition for rulemaking to add nutrient limits to the requirements for secondary treatment discussed in Chapter 1.0 is an example. In many States, federal district court rulings on TMDL lawsuits have resulted in requirements for State environmental agencies and the EPA to develop and implement TMDLs and have established other conditions for water quality protection that impact wastewater dischargers. The Friends of the Wild Swan v. EPA is an example of such a lawsuit. Other states have established "no net increase" policies that impact dischargers to streams that where TMDLs are pending. In the notable case of Friends of Pinto Creek v. EPA and Carlota Copper Company (2007), a precedent has been set for new discharges to streams with a TMDL that may delay discharge permitting. In Florida, Earthjustice filed a notice of intent to sue EPA on the lack of state nutrient standards and EPA has issued a determination that should Florida not make progress, then EPA will step in to set nutrient standards. A brief summary of each of these legal activities is presented in the following sections.

2.13.1 Friends of the Wild Swan v. EPA

Founded in 1987, Friends of the Wild Swan is a group focused on preserving the water quality, fish, wildlife habitat, road-less areas and wilderness in the Swan Valley in Montana. The stated objectives of this group are to accomplish water quality goals through: 1) administrative processes; 2) public education; 3) research; and 4) litigation. Protecting the Swan River and lake aquatic and terrestrial ecosystems is a primary focus.

Friends of the Wild Swan filed a lawsuit in federal district court to prevent increased pollutant loadings to water quality impaired streams in Montana. The Montana federal district court order in Friends of the Wild Swan, commonly referred to as the Judge Malloy decision, reads as follows:

"On September 21, 2000, a U.S. District Judge issued an order stating that until all necessary total maximum daily loads (TMDLs) under Section 303(d) of the Clean Water Act are established for a particular water quality limited segment, the State is not to issue any new permits or increase permitted discharges under the MPDES program. The order was issued in the lawsuit Friends of the Wild Swan v. U.S. EPA, et al., CV 97-35-M-DWM, District of Montana, Missoula Division. The DEQ finds that the issuance of this proposed permit does not conflict with the order because it is not a new permit."

The Montana Department of Environmental Quality must review each proposed NPDES discharge permit for compliance with the Malloy decision and demonstrate that the permit does not authorize any new or increased discharge of pollutants

2.13.2 No Net Increase Policies and Regulations

Some states have adopted no net increase policies and regulations to control loadings to impaired waterbodies when TMDLs are in progress, or have not been completed. This can create a constrained situation for those seeking new discharge permits and for existing dischargers seeking increases in permitted loadings. The issuance of NPDES permits prior to the completion of a TMDL is typically set in such a manner as to not cause or contribute to the existing impairment. Typically, this results in end-of-pipe criteria defined in such cases as not causing or contributing.

An example is the new "no net increase" regulation adopted by the state of Idaho. Idaho statue 39.3610 establishes for waters not requiring a TMDL that actions be taken "to prohibit further impairment of the designated or existing beneficial uses" (Idaho Code, 2009). The actions may include "changes in permitted discharges from point sources on the water body or to the best management practices for nonpoint sources within the watershed" (Idaho Code, 2009). Similar language is included in the Idaho Administrative Code 58.01.02 and adds "the total load remains constant or decreases within the watershed" (IDAPA, 2009a). Similar language is also included in the Idaho Administrative Code 16.01.02 (IDAPA, 2009b). The intent of the law and rule is to maintain or improve water quality conditions. The Idaho Department of Environmental Quality has interpreted this direction in policy PM98-2: Policy for No-Net Increase (IDEQ, 1998). Idaho DEQ developed regulations to implement the law and in doing so translated that the total load must remain constant, or decrease within the watershed, to meet the concept of no-net increase. This was a simple way of expressing the "idea in policy to hold the line on pollutant loads" (Essig and Grafe, 2005).

As Idaho DEQ has implemented its policy, the no-net increase terminology has become attached to the law and rules and DEQ has used the term "No Net Increase Rule" when referring to the Idaho Administrative Code in response to comments on water quality assessments and TMDLs (IDEQ, 2009). The policy has provided a means to start discussions on water quality trading. For example, on the lower Boise River, a framework has been developed for trading to meet the requirement for no-net increase in sediment and phosphorus. This watershed includes a mixture of stakeholders including agriculture, municipalities, industries, and other interests resulting in diverse interests and multiple possibilities for trading.

2.13.3 Friends of Pinto Creek v the United States Environmental Protection Agency and Carlota Copper Company

The United States Court of Appeals for the Ninth Circuit issued a ruling in the case of Friends of Pinto Creek v. EPA and Carlota Copper Company on October 4, 2007. This ruling has had implications for permitting new discharges and renewal of discharge permits with increased loadings in situations where a total maximum daily load (TMDL) is being prepared for the receiving waters. In order for a new discharge to be permitted, the Carlota Copper decision indicates that a TMDL needs to be completed and compliance schedules issued for other point source discharges.

2.13.3.1 Pinto Creek Background

The Carlota Copper situation on Pinto Creek is summarized from the original court case as follows:

> *"Pinto Creek is a desert river located near Miami, Arizona, approximately 60 miles east of Phoenix. It has been listed by the American Rivers Organization as one of the country's most endangered rivers due to threats from proposed mining operations. Pinto Creek and its riparian environs are home to a variety of fish, birds, and other wildlife,*

some of which are specially protected. Due to excessive copper contamination from historical mining activities in the region, Pinto Creek is included on Arizona's list of impaired waters under § 303(d) of the Clean Water Act, 33 U.S.C. § 1313(d), as a water quality limited stream due to non-attainment of water quality standards for dissolved copper.

Carlota proposed to construct and operate an open-pit copper mine and processing facility approximately six miles west of Miami, Arizona, covering over 3000 acres while extracting about 100 million tons of ore. Part of the operation plan includes constructing diversion channels for Pinto Creek to route the stream around the mine, as well as groundwater cutoff walls to block the flow of groundwater into the mine." – [From the original court case]

After trying, and failing, to get a general stormwater permit, Carlota applied for an individual NPDES permit under Section 402 of the CWA. After public comments were received on the draft NPDES permit, the EPA approved the permit with two conditions: 1) a requirement for additional groundwater discharges to augment the Pinto Creek stream flow; and 2) a requirement that Carlota perform remediation measures concerning sources of copper loading from an upstream inactive mine site. This is the permit that the Friends of Pinto Creek petitioned the Ninth Circuit Court to determine whether the EPA properly issued a National Pollution Discharge Elimination System permit under the Clean Water Act to Carlota Copper Company. The Ninth Circuit Court ruled in favor of the Friends of Pinto Creek and vacated the permit.

In its ruling, the Ninth Circuit Court cited 40CFR Section 122.4 and stated the following:

"No permit may be issued: (i) To a new source or a new discharger if the discharge from its construction or operation will cause or contribute to the violation of water quality standards. The owner or operator of a new source or new discharger proposing to discharge into a water segment which does not meet applicable water quality standards or is not expected to meet those standards . . . and for which the State or interstate agency has performed a pollutants load allocation for the pollutant to be discharged, must demonstrate, before the close of the public comment period, that: (1) There are sufficient remaining pollutant load allocations to allow for the discharge; and (2) The existing dischargers into that segment are subject to compliance schedules designed to bring the segment into compliance with applicable water quality standards."

According to the Ninth Circuit Court ruling, "*the first sentence of the regulation is very clear that no permit may be issued to a new discharger if the discharge will contribute to the violation of water quality standards*." That is, there is no provision for immediate offset. But the judge goes on to say that there are two exceptions to this rule, and because of these exceptions, the argument that this would lead to a blanket moratorium for new dischargers in impaired waters lacks merit. The judge ruled that the two exceptions, which must be read and interpreted together, are that there are sufficient remaining pollutant load allocations to allow for the discharge, as calculated in a TMDL, and that a compliance schedule be designed to bring the segment into compliance with applicable water quality standards. Additionally, the compliance schedule must include all point sources, not just permitted point sources.

2.13.3.2 Appeal and Potential Ramifications

The Ninth Circuit Court decision was appealed to the US Supreme Court by Carlota. The National Association of Clean Water Agencies (NACWA), along with several other non-governmental organizations (NGO's) filed a statement of interest of Amici Curiae (Brief) on behalf of Carlota. However, as this case was not heard by the US Supreme Court, the judgment of the 9th Circuit Court stands as law in the western states and is a basis for citing precedent in the rest of the country.

The potential implication of the Ninth Circuit Court decision is that a discharger to an impaired waterbody may have to wait until a TMDL is completed to receive a new permit with increased loading, or to permit a new discharge. Water quality standard compliance schedules must also be completed, and not just for permitted point sources, but also for non-permitted point sources. The compliance schedule requirement contains echoes of the proposed changes to the Clean Water Act in 2000.

Several points were raised in the NACWA Brief on the ramifications of the Carlota Copper ruling. The ruling could potentially delay new NPDES permit where a TMDL is not completed because it can take 13 years or more to complete a TMDL, resulting in a de facto moratorium on new growth. While the ruling does not specifically exclude off-setting pollution, it does exclude it without first performing a pollutants load allocation for the pollutant to be discharged and demonstrate, before the close of the public comment period, that there are sufficient remaining pollutant load allocations to allow for the discharge. This ruling on offsetting, according to the NACWA Brief, contradicts several other court precedents, and several EPA manuals on water quality trading.

There are many new issues raised by Carlota Copper ruling that will have to be addressed either by the EPA, U.S. legislature, or through further case law. Concerns include how to address non-permitted point sources, whether they are in violation of the Clean Water Act, and how they are to be found. Another concern is the use of phased TMDLs in an adaptive management approach to a water quality plan and whether this approach meets the conditions in the Clean Water Act as suggested by Carlota Copper decision. This view on water quality offsets may inhibit Water quality trading processes in the future.

As posed by Karl Blankenship in the Bay Journal News Service (2009), if NGOs (Environmental Groups) petitioned and this ruling was applied to the Chesapeake Bay, where there is no TMDL for the Bay itself, growth (in one of the fastest growing areas in the nation) would come to a complete halt until a TMDL could be completed. Although there are currently 'Tributary Strategies' in the watershed, there is concern that these would not qualify as appropriate compliance schedules to meet the other requirements suggested by the Carlota Copper decision.

2.13.4 Florida Wildlife Federation, Sierra Club, Conservancy of Southwest Florida, Environmental Confederation of Southwest Florida, and St. John's Riverkeeper v. EPA

In a letter dated April 29, 2008, Earthjustice (2008) stated their intention to sue the United States Environmental Protection Agency for not setting numeric nutrient criteria for Florida as outlined in section 303(c)(4)(B) of the Clean Water Act (CWA). The case was filed in the U.S. District Court, Northern District of Florida involved the Florida Wildlife Federation, Sierra Club, Conservancy of Southwest Florida, Environmental Confederation of Southwest Florida, and St. John's Riverkeeper (2008).

In a letter to the Florida Department of Environmental Protection dated January 14, 2009, Ben Grumbles, Assistant Administrator for the EPA (EPA, 2009d), issued an official determination that pursuant to the CWA section 303(c)(4)(B), new or revised nutrient water quality standards are necessary for Florida to meet the requirements of the CWA. Although Florida has taken steps to control nutrient enrichment, EPA concluded that Florida's narrative nutrient criterion is not sufficient to protect designated uses and numeric criteria are necessary to comply with the CWA.

Florida's nutrient control efforts have included:

- Adoption of nutrient specific narrative criterion and assessment procedures through its Impaired Waters Rule (IWR),
- Promoting watershed management plans through the Basin Management Action Plans (BMAPs), and
- Enactment of additional laws and programs for point and nonpoint source control, including the Grizzle-Figg Act of 1990.

The Florida narrative criterion for nutrients states that, "in no case shall nutrient concentrations of a body of water be altered so as to cause an imbalance in natural populations of aquatic flora or fauna." Determination of this concentration for waters in the state can be a lengthy process. The EPA determination states that numeric nutrient criteria "would enhance effectiveness of NPDES permits in protecting designated uses and enable Florida permit writers to derive effluent limitations without the resource intensive and burdensome process of conducting site specific analyses to determine the appropriate numeric target value."

The EPA letter of determination also states that numeric nutrient criteria would have similar effects on TMDL development and would shorten the delay that occurs with the state's Impaired Waters Rule implementation. Florida's Basin Management Action Plans utilizes stakeholder involvement in the development of TMDLs. The Grizzle-Figg Act requires effluent limits of 5 mg/L BOD, 5 mg/L suspended solids, 3 mg/L total nitrogen, and 1 mg/L total phosphorus (5/5/3/1) for all domestic wastewater treatment facilities in the Tampa Bay area.

The EPA letter of determination states that despite Florida's actions to control nutrient pollution, analysis of United States Geological Survey (USGS) monitoring data shows that no significant improvements have occurred since 1980. According to STORET data, the concentration of total phosphorus remains constant with a mean of 0.15 mg/L and total nitrogen averages about 1.4 mg/L. The 2002 CWA section 303(d) list reports that over 60% (or 550 of 823 waters) are impaired for nutrient causes.

In its determination letter, the EPA maintains that numeric nutrient criteria are necessary to meet the requirements of the CWA and to protect Florida's designated uses, and that Florida must *"conduct case-by-case assessments to determine if an imbalance in flora or fauna exists for waters below the IWR impairment thresholds."*

The remedy for the situation is described in the letter of determination: "EPA will move forward to develop federal proposed regulations setting forth numeric nutrient criteria for Florida and expects that these criteria will be developed in a manner that ensures that there will be no imbalance in natural populations of flora and fauna in Florida waters."

Following the January 14, 2009 letter to the Florida Department of Environmental Protection EPA expected that data collection would be completed by March 2009 by the state of Florida and that it would take an estimated six months to complete data analysis. EPA expected to develop criteria values for lakes and flowing waters within 12 months and for estuaries and coastal water within 24 months. In addition, the EPA stated in its remedy that should Florida adopt and EPA accept revised water quality standards, then EPA would not need to issue federal standards: "*approves new or revised water quality standards that sufficiently address this determination before EPA promulgates federal water quality standards, EPA would no longer be obligated to promulgate federal water quality standards.*"

2.13.5 EPA Proposed Water Quality Standards for Florida

On January 26, 2010 EPA proposed water quality standards for lakes and flowing waters in the state of Florida. The EPA summarizes the proposed rule as follows:

> *The Environmental Protection Agency (EPA) is proposing numeric nutrient water quality criteria to protect aquatic life in lakes and flowing waters, including canals, within the State of Florida and proposing regulations to establish a framework for Florida to develop ''restoration standards'' for impaired waters. On January 14, 2009, EPA made a determination under section 303(c)(4)(B) of the Clean Water Act (''CWA'' or ''the Act'') that numeric nutrient water quality criteria for lakes and flowing waters and for estuaries and coastal waters are necessary for the State of Florida to meet the requirements of CWA section 303(c). Section 303(c)(4) of the CWA requires the Administrator to promptly prepare and publish proposed regulations setting forth new or revised water quality standards (''WQS'' or ''standards'') when the Administrator, or an authorized delegate of the Administrator, determines that such new or revised WQS are necessary to meet requirements of the Act. This proposed rule fulfills EPA's obligation under section 303(c)(4) of the CWA to promptly propose criteria for Florida's lakes and flowing waters.*

EPA is proposing four water body types for the State of Florida upon which to base nutrient standards: lakes, streams, springs and clear streams, and canals in south Florida. EPA's proposed rule includes nutrient criteria for both in-stream protection values and downstream protection values (EPA, 2010). The proposed rule would:

- Set total nitrogen and total phosphorus limits for the protection of lakes, streams, and canals (in-stream protection values).
- Set a second set of limits for total nitrogen and total phosphorus for waters that flow into lakes and estuaries to ensure protection of those downstream waters (downstream protection values or DPVs).

The more stringent of the two criteria would apply for each waterbody. More stringent criteria in an upstream waterbody are intended to protect aquatic life in the downstream waterbody such lakes and estuaries. Based on the data, the DPV will likely be lower than the in-stream protection value for many streams in Florida (FWEA, 2010).

For rivers and streams in Florida, EPA has proposed in-stream protection values as numeric nutrient criteria based on four watershed regions described in Table 2-6.

Table 2-6. Summary of EPA's Proposed Nutrient Criteria for Florida.

Nutrient Watershed Region	In-stream Protection Value Criteria	
	TN (mg/L)[a]	TP (mg/L)[a]
Panhandle [b]	0.043	0.043
Bone Valley [c]	1.798	0.739
Peninsula [d]	1.205	0.107
North Central [e]	1.479	0.359

[a] Concentration values are based on annual geometric mean not to be surpassed more than once in a three-year period. In addition, the long term average of annual geometric mean values shall not surpass the listed concentration values. (Duration = annual; Frequency = not to be surpassed more than once in a three-year period or as a long-term average).
[b] Panhandle region includes the following watersheds: Perdido Bay Watershed, Pensacola Bay Watershed, Choctawhatchee Bay Watershed, St. Andrew Bay Watershed, Apalachicola Bay Watershed, Apalachee Bay Watershed, and Econfina/Steinhatchee Coastal Drainage Area.
[c] Bone Valley region includes the following watersheds: Tampa Bay Watershed, Sarasota Bay Watershed, and Charlotte Harbor Watershed.
[d] Peninsula region includes the following watersheds: Waccasassa Coastal Drainage Area, Withlacoochee Coastal Drainage Area, Crystal/Pithlachascotee Coastal Drainage Area, Indian River Watershed, Caloosahatchee River Watershed, St. Lucie Watershed, Kissimmee River Watershed, St. John's River Watershed, Daytona/St. Augustine Coastal Drainage Area, Nassau Coastal Drainage Area, and St. Mary's River Watershed.
[e] North Central region includes the Suwannee River Watershed.

In the proposed rule, EPA acknowledges the important water resource role of clear streams and springs to the people of Florida and the anthropogenic effects that have caused degradation to these resources. The numeric nutrient criteria proposed by EPA for springs and clear streams (< 40 PCU) is written as follows in the proposed rule:

> *Nitrate (NO3)+Nitrite (NO2) shall not surpass a concentration of 0.35 mg/L as an annual geometric mean more than once in a three-year period, nor surpassed as a longterm average of annual geometric mean values.*

EPA also dictates that TN and TP criteria for streams on a watershed basis also apply to clear streams.

In the proposed rule, EPA describes the diversity of canals and how they have changed ecosystems and hydrology in Florida. EPA proposes numeric nutrient criteria for canals classified as Class III waters under Florida Administrative Code (Rule 62-302.400). EPA notes that that proposed criteria would not apply for TP in canals within the Everglades Protection Area (EvPA) as a TP criterion of 0.010 mg/L currently applies to this area.

2.13.5.1 Downstream Protection Values

In an effort to protect downstream reaches, EPA presents numerous criteria tables with proposed Downstream Protection Values (DPVs) for estuaries that include an annual protective TN and TP loading value for the corresponding estuary. EPA has proposed a methodology for calculating criteria for streams that flow into lakes and estuaries. The Downstream Protection Values are based on annual TN and TP load estimates based on assimilation capacity of the lake or estuary. Three alternatives are proposed for determining the DPV:

- Use EPA's Downstream Protection Values
- Use EPA's Downstream Protection Values methodology but with redistribution of the load among each of the tributaries of an estuary
- Use another defensible alternative quantitative methodology

EPA used the following method to develop numeric nutrient criteria for streams with regard to protecting downstream waterbodies:

1. Protective Load: Determination of the average annual nutrient load that can be delivered to an estuary without impairing designated uses.
2. Downstream Protection Values: Determination of nutrient concentrations in an estuary watershed that result in a nutrient loading that does not exceed the protective load. EPA used the SPAtially Reference Regressions on Watershed (SPARROW) model to determine these concentrations.

The Florida DEP has expressed concerns with this approach and stated "EPA's stream criteria for protection of downstream estuaries are not scientifically valid" (FDEP, 2010a). Concerns expressed by Florida DEP include the following (FDEP, 2010b):

- *The method "does not include settling terms for streams leading to the lake"*
- *"Other sources of nutrients to lakes like groundwater and atmospheric deposition are not included"*
- *"Model estimates of in-stream nitrogen losses are too low"*
- *"Method for establishing acceptable loads to estuaries not based on cause/effect relationships"*

Various water resource agencies and groups (such as the Florida Stormwater Association, Florida Water Environment Association, and Tampa Bay Nitrogen Management Consortium) have expressed similar concerns (FSA, 2010, FWEA, 2010, NMC, 2010). "The rule derived criteria based on a statistical distribution of nutrients present in waters judged to be in a healthy biological condition" is the "subject of considerable scientific debase as to its validity" (Arnold, 2010). It has also been observed that EPA is seeking input on this method, as highlighted in the following statement from the Federal Register "EPA is interested in obtaining feedback at this time on this systematic and scientific approach" (EPA, 2010).

2.13.5.2 Alternative Regulatory Approaches and Implementation Mechanisms

EPA's proposed Florida nutrient criteria includes a section called "*Alternative Regulatory Approaches and Implementation Mechanisms*" that presents a discussion of several tools for implementing nutrient control requirements, including the following:

- Water Quality Variances
- Site Specific Alternative Criteria (SSAC)

- Compliance Schedules
- Restoration Water Quality Standards

EPA's newly proposed Restoration Standard approach for Florida introduces a new tool which appears similar in some characteristics to a water quality variance in that full compliance can be extended by phasing for eventual attainment of water quality standards. EPA has framed Restoration Standards to consider a broad range of watershed management issues, including nonpoint source controls, and innovative and flexible approaches. The Restoration Standards concept emphasizes beneficial use attainment in the watershed in phased milestone steps. Restoration Standards would carry a substantial burden in building the justification to demonstrate that standards are not attainable and in defining interim milestones and beneficial uses, perhaps through use attainability analysis (UAA). In the proposed rule, EPA has specifically requested input on defining what constitutes "maximum feasible progress" in the phased implementation of a Restoration Standard.

2.14 Antidegradation Regulations

Antidegradation regulations are designed to provide a decision-making framework about protecting existing high water quality where the water exceeds the levels necessary to support propagation of fish/shellfish/wildlife and recreation in or on the water, and to ensure protection of existing uses for all waters of the U.S.

Requirements vary from state-to-state and emphasize a constituent-by-constituent and waterbody-by-waterbody analysis. Antidegradation policies or regulations may create conditions that limit nutrient discharges as flows and loadings to wastewater facilities increase with community growth. Wastewater utilities may need to incorporate additional levels of treatment technology beyond that required in an existing discharge permit in order to preserve 'room for growth.'

This section discusses antidegradation and summarizes the origins of regulations and their application by states. Historically, states have implemented antidegradation policy from the perspective of "*How much pollution can be incorporated into a waterbody before it loses one of its designated uses?*" More recently, EPA (2007b) has emphasized moving states to a focus on current water quality and how can it be protected from any further degradation.

2.14.1 Antidegradation History

The first Antidegradation policy was issued by the Department of Interior in February 1968 and pre-dates the Clean Water Act. Antidegradation was included in the first EPA Water Quality Standards Regulation (40 CFR 130.17,40 F.R. 55340-41, November 28, 1975) as the beginning of Clean Water Act water quality requirements and was re-issued as part of the current regulations (48 F.R. 51400, 40 CFR 131.12, November 8, 1983).

2.14.1.1 Clean Water Act

The Clean Water Act includes antidegradation policy based on the goal to "... restore and maintain the chemical, physical and biological integrity of the Nation's waters" and specifically in 40 CFR Section 131.12 which is summarized on the next page:

40 CFR § 131.12 Antidegradation policy.

(a) The State shall develop and adopt a statewide antidegradation policy and identify the methods for implementing such policy pursuant to this subpart. The antidegradation policy and implementation methods shall, at a minimum, be consistent with the following:

(1) Existing instream water uses and the level of water quality necessary to protect the existing uses shall be maintained and protected.

(2) Where the quality of the waters exceed levels necessary to support propagation of fish, shellfish, and wildlife and recreation in and on the water, that quality shall be maintained and protected unless the State finds, after full satisfaction of the intergovernmental coordination and public participation provisions of the State's continuing planning process, that allowing lower water quality is necessary to accommodate important economic or social development in the area in which the waters are located. In allowing such degradation or lower water quality, the State shall assure water quality adequate to protect existing uses fully. Further, the State shall assure that there shall be achieved the highest statutory and regulatory requirements for all new and existing point sources and all cost-effective and reasonable best management practices for nonpoint source control.

(3) Where high quality waters constitute an outstanding National resource, such as waters of National and State parks and wildlife refuges and waters of exceptional recreational or ecological significance, that water quality shall be maintained and protected.

(4) In those cases where potential water quality impairment associated with a thermal discharge is involved, the antidegradation policy and implementing method shall be consistent with section 316 of the Act.

The Clean Water Act incorporates antidegradation policy in a 1987 amendment:

a 1987 amendment codified in section 303(d)(4)(B) requiring satisfaction of antidegradation requirements before making certain changes in NPDES permits; and the 1990 Great Lakes Critical Programs Act codified in CWA section 118(c)(2) requiring EPA to publish Great Lakes water quality guidance including antidegradation policies and implementation procedures.

Tiers of protection are defined in the Clean Water Act. For states, the course of action for Tier 1 protecting existing uses and Tier 3 outstanding natural resources, the requirements are clear. Protection of Tier 2 waters for "fishable/swimmable" uses is less clear. Protecting high quality Tier 2 waters can be more complex because it involves an antidegradation review process that includes components such as an alternatives analysis and a social and/or economic analysis.

Section 131.13(a)(1), or "Tier 1," protecting "existing uses," provides the absolute floor of water quality in all waters of the United States. This paragraph applies a minimum level of protection to all waters.

Section 131.12(a)(2), or "Tier 2," applies to waters whose quality exceeds that necessary to protect the section 101(a)(2) goals of the Act. In this case, water quality may not be lowered to less than the level necessary to fully protect the "fishable/swimmable" uses and other existing uses and may be lowered even to those levels only after following all the provisions described in section 131.12(a)(2).

Section 131.12(a)(3), or "Tier 3," applies to Outstanding National Resource Waters (ONRW) where the ordinary use classifications and supporting criteria may not be sufficient or appropriate. As described in the preamble to the Water Quality Standards Regulation. "States may allow some limited activities which result in temporary and short-term changes in water quality," but such changes in water quality should not impact existing uses or alter the essential character or special use that makes the water an ONRW.

2.14.2 Application of Antidegradation by States

For states, antidegradation policy can be a controlling factor for nutrients even if a state does not have specific numeric water quality standards for nitrogen or phosphorus. This can occur by linking nutrients to a narrative water quality standard for an offensive condition, or through the basic designated beneficial uses of the waterbody.

2.14.2.1 Threshold for 'de minimis' Impact

States have a 'de minimis' threshold on the loss of assimilative capacity that would be allowed to occur without triggering an antidegradation review. States' laws differ on when antidegradation reviews are required and on which waters. For some states, the 'de minimis' threshold is 1%. Generally, states have an upper limit on the amount of degradation allowed, such as 10%, providing that it does not lead to a waterbody losing a designated beneficial use.

2.14.2.2 Socioeconomic Impacts

Economic criteria are used to evaluate whether degradation of surface waters is warranted with a pollution control project. The EPA's "Interim Economic Guidance for Water Quality Standards" provides a process for consideration of these criteria (EPA, 1995). The main concerns are "***whether the pollution controls needed to maintain the high-quality water will interfere with the proposed development***," and "***if so, then the review must show that the development would be an important economic and social one***" or financial feasibility and socioeconomic impacts.

Financial feasibility considers the annualized cost of the pollution controls, who pays for it, and whether the cost imposed is reasonable. For example a local public agency must identify the cost of pollution control; develop the method to fund it (i.e. municipal bonds backed by user fees); calculate the increase in annual household costs; and then screen the pollution measure by comparing the increase in household costs to annual household incomes.

If the pollution control measure is financially infeasible, then the entity evaluates the economic importance of the reduction in water quality using criteria such as: unemployment rate; impact on community development potential; impact on property values; and impacts to tax revenues. The EPA's guidance notes that "***There are no economic ratios per se that determine whether a development would be considered important***" and that "***The term important is intended to convey a general concept regarding the level of social and economic development used to justify a change in high-quality waters.***" Put simply, economic importance is addressed in an ad hoc manner with a relative comparison of all economic criteria and justification for degradation and will likely vary on a case-by-case basis.

2.14.2.3 Burden of Proof

The burden of proof to demonstrate that a discharge will not contribute to the degradation of the waterbody generally falls to the discharge permit applicant. Determining whether a discharge will lead to degradation can be challenging. Each state classifies its own waterbodies into one of three or four tiers, and each tier requires a different level of scrutiny. Each state has varying procedures for antidegradation review ranging from a cursory study, to more in-depth analysis. If a permit is challenged by the public, there could be further requirements under a state's antidegradation policy. A complicating factor with respect to nutrients in the absence of numeric nutrient standards is that a water quality study and modeling effort might be required to determine the level of degradation for a specific increase in phosphorus or nitrogen discharge.

2.14.3 State by State

The following paragraphs summarize the antidegradation policies and regulations as they are applied in individual states. In some cases, case studies illustrate where antidegradation policy has come into play.

2.14.3.1 Colorado

In Colorado, the state antidegradation review consists of the following steps: applicability, significance determination, necessity of degradation determination, and protection of existing uses (CDPHE, 2008). For bioaccumulative toxic pollutants, degradation is not significant if the new/increased loading is less than 10% of the existing total load, provided that the cumulative increased loading is not more than 10% of the baseline total load. For other pollutants, such as nutrients, degradation is not significant if: (a) the low flow dilution ratio is 100 to 1 or more, or (b) the activity will consume, after mixing, less than 15% (cumulatively) of the baseline assimilative capacity (default baseline is as of September 30, 2000), or (c) the activity will result in only temporary or short term changes in water quality.

2.14.3.2 Georgia

Georgia's antidegradation policy influenced permitting the permitting of a wastewater treatment plant and potential water quality offset. It is only through the building of facilities capable of levels of treatment beyond their permits that facilities can trade water quality credits. If antidegradation policy is written is such a way as to force all permit holders to achieve the maximum level of treatment possible, there will be no room for water quality trading.

A possible conflict between the goal of increased water quality trading and antidegradation policy arose in Gwinnett County where the F. Wayne Hill Water Resources Center had upgraded its facilities with state-of-the art ultrafiltration technology and was issued a permit to discharge. The plant was able to achieve much greater levels of treatment than was allowed by the permit that was issued by the State of Georgia. The permit was challenged on the basis that the permit violated the State's antidegradation policy because the policy requires permitees to utilize the "***highest and best [level of treatment] practicable under existing technology.***" Since the plant was capable of removing more pollution than the permit required, the permit discharge limits were tightened to match the level of treatment capable by the facility. The state's antidegradation policy has since been changed to eliminate this sentence.

2.14.3.3 Idaho

The Idaho Conservation League provided notice of intent to sue EPA regarding the inaction of the State of Idaho to develop an antidegradation implementation plan (Advocates for the West, 2009). The assertion is that because Idaho's water quality standard lack an implementation plan, EPA should not approve any water quality standards until a plan is developed. If the State of Idaho does not develop a plan, then EPA should develop a plan for the state. The notice of intent includes the argument that the antidegradation policy requires state standards be sufficient to maintain existing beneficial uses. Therefore, without the policy and plan, it is impossible to know if appropriate water quality standards are being set.

The claim also argues that three tiers of waters need to be defined as part of the antidegradation policy and Idaho has also failed to identify any methods to implement a policy that relates the tiers to protecting water quality. The suit claims that Idaho and EPA have failed to follow the Clean Water Act requirements regarding antidegradation implementation plans and the setting of water quality standards.

The Idaho Conservation League further used this argument to petition the Environmental Appeals Board concerning the issuance of an NPDES permit for the City of Twin Falls Wastewater Treatment Plant (EPA EAB, 2009e). Included in the petition is the assertion that Idaho does not have a lawful antidegradation policy and implementation plan.

2.14.3.4 Illinois

The Illinois Town of New Lenox and the Illinois Environmental Protection Agency (IEPA) sought to permit a wastewater treatment plant expansion and Environmental groups petitioned against issuing the permit (IEPA, 2008). In a court ruling by the 3rd District Appellate Court of Illinois, filed October 7, 2008 found that even though there were no explicit nitrogen or phosphorus standards, because nutrient have been found to be linked to BOD and pH, nitrogen and phosphorus *must* be included in the antidegradation study and that the City and IEPA did not demonstrate that lowering of water quality in the waterbody was necessary to accomplish important socioeconomic development in the area.

2.14.3.5 Iowa

There is currently a bill before the Iowa State Senate proposing state antidegradation legislation (Iowa, 2009). The bill requires the Department of Natural Resources (IDNR) to establish and administer a three-tiered antidegradation policy which is in accordance with the Clean Water Act. For the Tier 2 review and compliance requirements, the IDNR will consider alternatives with costs less than 110% of the base cost of the pollution control measures for the discharge, as economically feasible. Alternatives with costs greater than 110% may be considered if the alternative is proven to produce a substantial improvement in the resulting discharge and significantly improve and protect water quality. Another aspect of Iowa's proposed legislation that differentiates it from other states' antidegradation policies is the procedure that is laid out for the State and the public to nominate and add to the State's list of 'outstanding' waters.

2.14.3.6 Kentucky

The U.S. Court of Appeals for the Sixth Circuit issued an opinion on September 3, 2008 in the case of Kentucky Waterways Alliance v. Johnson, reversing and remanding in part EPA's approval of Kentucky's antidegradation rules. The Plaintiffs in the case, Kentucky Waterways Alliance, Sierra Club Cumberland Chapter, Kentuckians for the Commonwealth, and Floyds Fork Environmental Association, objected to Kentucky's antidegradation policy. The court ruled that EPA's approval of 5 of the 6 exceptions to the requirement of justifying a lowering of water quality of high quality waters was "arbitrary and capricious" because EPA never required Kentucky to prove that the multiple exceptions contained in Kentucky's rules would cause only insignificant, or "de minimus," degradation of the state's rivers, lakes and streams. One of the exemptions in question was a blanket exemption of coal mining discharges from antidegradation review, even though the state offered a letter of commitment to the EPA stating that that it would interpret its regulation to require antidegradation review.

2.14.3.7 Maryland

An alternatives analysis must be completed as part of all antidegradation reviews where antidegradation reviews are required (Maryland, 2009). However, the social and economic justification aspect of the antidegradation review is required only if the result of the discharge would be that assimilative capacity is cumulatively reduced (all sources) by more than 25% from the baseline water quality determined when the water body was listed as Tier 2.

2.14.3.8 Massachusetts

In Massachusetts, even though there are explicit nutrient criteria for several waterbodies in the state, the State's antidegradation policy clearly states that when a waterbody has no explicit nutrient criteria, the excessive growth of weeds and algae impairing designated uses will be used to guide antidegradation reviews (MDEP, 2008). This eutrophication is explicitly described as being caused by total mass loading of nutrients, nutrient ratios, nutrient recycling, and other factors.

2.14.3.9 Missouri

Missouri emphasizes an antidegradation focus on pollutant-by-pollutant and waterbody by waterbody analysis (MDNR, 2008). Degradation of assimilative capacity may be allowed if it is considered minimal degradation, or if it is justified in accordance with an antidegradation review. Degradation is considered minimal if the reduction of assimilative capacity as a result of the new or proposed loading (i.e., event-specific) is less than 10%, and the loss of assimilative capacity as a result of cumulative degradation is less than 20%. "Cumulative Degradation" is the reduction of a segment's assimilative capacity from separate discharges approved by the department following the establishment of the water's existing water quality.

2.14.3.10 Montana

The Montana Department of Environmental Quality administers state non-degradation rules to protect surface water and groundwater quality (17.30.701 et seq. Administrative Rules of Montana (ARM)) (MCA, 2009). Nondegradation determinations are typically associated with an effluent discharge mixing zone where dilution of a discharge may occur. Some nondegradation limits are set at definite concentrations called a trigger, or at a percentage of the lowest applicable water quality standard. Other nondegradation limits are qualitative, such as those for nitrogen and phosphorus in surface water.

Montana exempts certain activities (75-5-317, MCA) exempts certain activities from the nondegradation requirements (i.e. automatically classifies them as "nonsignificant"). Exemptions are based on the activities low potential for harm to human health and the environment (75-5-301(5)(c), MCA). The nondegradation rules only apply to "new or increased sources" as of April 29, 1993 (ARM 17.30.702(16)). This clause exempts discharges that were existing or permitted, authorized or approved prior to April 29, 1993. These exemptions do not apply to the state water quality standards, which include the human health and aquatic life standards listed in the surface and ground water standards.

The Montana Water Quality Act authorizes the issuance of point source discharge permits on a listed water body pending completion of a TMDL provided that: 1) the discharge is in compliance with the provisions of 75-5-303, MCA (nondegradation policy); and 2) the discharge will not cause a decline in the water quality of the parameter for which the waterbody is listed [75-5-303(10), MCA].

On September 21, 2000, a US District Judge issued an order (Friends of the Wild Swan vs. US EPA *et al*, CV 97-35-M-DWM, District of Montana, Missoula Division) stating that until all necessary total maximum daily loads under Section 303(d) of the Clean Water Act are established for a particular water quality limited segment, the State is not to issue any new permits or increase permitted discharges under the MPDES program.

2.14.3.11 New Mexico

A phosphorus TMDL of Rio Hondo in New Mexico is an example of the state's antidegradation policy (NMED, 2005). Through loading calculations, it was determined that there was 1.47 lbs/day of phosphorus available for wasteload allocations. However, since the current wasteload from point sources was calculated to be 1.0 lbs/day, the State of New Mexico, based on the State's antidegradation policy, limited the wasteload allocation to 1.0 lbs/day. This is a clear case of state antidegradation policy affecting a nutrient TMDL allocation and limiting future NPDES permit increases.

2.14.3.12 North Carolina

In North Carolina, each applicant for an NPDES permit, or NPDES permit expansion, to discharge treated waste must document an effort to consider non-discharge alternatives under the state's antidegradation policy (NCDENR, 2007).

2.14.3.13 Ohio

Ohio antidegradation policy is currently undergoing changes. A draft of the new regulations are dated from October 2008. One of the important updates of the policy is a revision of the definition of "best available demonstrated technology." This definition update will include new design criteria for nitrogen and phosphorus effluent limits.

2.14.3.14 Wyoming

Wyoming has a basic antidegradation policy that is covered by an agency policy rather than an adopted rule (WDEQ, 2007). The Wyoming antidegradation policy does not follow the EPA recommended three-tier system. Rather it follows the state's waterbody classification system instead.

2.15 Adaptive Management

In the August 2008 issue of the Water Environment Federation's journal, *Water Environment & Technology* (WE&T), Paul Freedman, Len Shabman, and Kenneth Reckhow contend that adaptive implementation can aid stakeholders in meeting water quality goals. The authors contend that the current approach of developing TMDLs may be out of date and that the U.S. is attempting to solve today's water quality issues using policies established 37 years ago (the original Clean Water Act of 1972).

The authors begin by discussing the uncertainty associated with the current TMDL development process, beginning with the listing of impaired waters, to the quantification of non-point sources, and to the efficacy of water quality controls, like best management practices. In addition, they state that these uncertainties often lead to long delays in the approval of new TMDLs.

To solve these problems, adaptive implementation, or "learning while doing," is suggested. With this approach, a project team would draft a TMDL and its implementation plan. This plan would then be implemented and the water of concern tested. The project team assesses the progress and revises the TMDL and implementation plan as needed within the CWA-delineated processes. The authors contend this approach would guarantee progress toward meeting water quality standards, including those for nutrients.

Freedman et al. explain that adaptive implementation is not the same as the standard TMDL implementation in which a TMDL is typically not revisited after a pollutant-reduction plan is in place, and that adaptive implementation would be most useful to watersheds with the most uncertainty. The authors state that adaptive implementation should not replace traditional TMDL implementation in many cases because of the potential time and effort involved in the adaptive implementation process.

Potential issues to be addressed with the adaptive implementation process include the present difficulty of revising a TMDL. In addition, the question remains of how adaptive implementation benefits a discharger that most likely had extremely low limits set in the TMDL. Perhaps a slight adjustment to the suggested adaptive implementation approach would be for dischargers to be involved early and adjust lines of thought as the data monitoring, analyses, and modeling is being performed to best understand the system and potential alternative allocations for the TMDL. TMDLs establish the maximum loadings that can be discharged into a waterbody and still meet the applicable standards. If the permitting authority chooses to implement the wasteload allocations developed in the TMDL in an adaptive approach, and this approach is consistent with the permitting regulations, EPA will support adaptive implementation.

2.16 References

40 CFR 131.12. Code of Federal Regulations – Title 40: Protection of Environment. Antidegradation Policy, December 2005.

Advocates for the West. 2009. Re: 60-Day Notice of Intent to Sue for Violations of the Clean Water Act Regarding Idaho's Antidegradation Policy Implementation Methods.

Arnold. 2010. EPA Proposes Numeric Nutrient Criteria for Florida's Surface Waters. Arnold&Porter LLP Adivsory, January 2010. http://www.arnoldporter.com/resources/documents/Advisory_EPAProposesNumericNutrientCriteriaforFlorida%E2%80%99sFlowingWaters_012510.pdf

Blankenship, Karl. Bay Journal News Service. 2009. http://www.bayjournalnewsservice.com/Desert.html

Clean Water. 2008. Update on Rule Revisions: Intent and Context. Clean Water From Politics to Policy. July 24, 2008. http://www.lwm-info.org/vertical/Sites/%7B92F7D640-E25A-4317-90AD-4976378A8F8D%7D/uploads/%7B19FD81F8-4642-40AE-8D00-9D123B07A454%7D.PDF

CDPHE. 2002. Nutrient Criteria Development Plan for Colorado. September 26, 2002, revised July 12, 2004. Colorado Department of Public Health & Environment, Water Quality Control Division, Denver, CO.

CDPHE. 2007. 2007 Progress Report and Update to the Colorado Nutrient Criteria Development Plan.

CDPHE. Water Quality Control Division. Regulation 31-The Basic Standards and Methodologies for Surface Water. May 31, 2008. http://www.cdphe.state.co.us/regulations/wqccregs/100231wqccbasicstandardsforsurfacewater.pdf

CDPHE. 2010a. Memorandum Re: Materials for November 16, 2009 Work Group Meeting, Nutrient Criteria Development. http://projects.ch2m.com/cwqf/Workgroups/Content/Nutrient_Criteria/Documents/General/Nutrients%20New%20Process%2011-16-09.pdf

CDPHE. 2010b. Nutrient Criteria for Rivers and Streams, Status/Update for Discussion February 9, 2010.

Colorado Water Quality Forum. 2010. Nutrient Criteria. http://projects.ch2m.com/cwqf/Workgroups/nutrient.asp

CSWEA. 2009a. Wisconsin Section Central States Water Environment Association, Committee Reports. May 20, 2009. http://www.cswea.org/wisconsin/committeereports/CSWEACommitteeReportsMay.20.2009Master.pdf

CSWEA. 2009b. Wisconsin Section Central States Water Environment Association, Committee Reports. August 12, 2009. http://www.cswea.org/wisconsin/committeereports/WI.CSWEACommitteeReportsAugus12,2009.pdf

DNR. 2008. Wisconsin Department of Natural Resources. 2008-2011 Triennial Standards Review Cycle. http://dnr.wi.gov/org/water/wm/wqs/tsr/documents/Topic_Descriptions.pdf

DNR. 2009. Wisconsin P Perspective. Wisconsin Department of Natural Resources. Russ Rasmussen, February 19, 2009. http://www.wwoa.org/files/Rassmussen%20Wi%20P%20Perspective.pdf

DNR. 2010a. Wisconsin Department of Natural Resources. Current Triennial Standards Review Projects. http://www.dnr.state.wi.us/org/water/wm/wqs/tsr/currenttsrprojects.htm#PWQC

DNR. 2010b. Wisconsin Department of Natural Resources. Wisconsin's 2010 Impaired Waters List. http://dnr.wi.gov/org/water/wm/wqs/303d/303d.html

Earthjustice letter to United States Environmental Protection Agency, "Intent to Sue the Administrator of EPA for Failure to Perform Its Mandatory Duty to Promulgate Numeric Nutrient Criteria for the State of Florida as Required by Section 303(c)(4)(B) of the Clean Water Act," April 29, 2008.

EPA, "Interim Economic Guidance Workbook," EPA 823-B-95-002, March 1995. http://www.epa.gov/waterscience/standards/econworkbook/

EPA. 1996. U.S. EPA NPDES Permit Writers' Manual, EPA-833-B-96-003, December 1996.

EPA, 2000a. Applicability of Ecoregion Criteria to Real World, EPA-822-B-00-002

EPA. 2000b., "Nutrient Criteria Technical Guidance Manual, Rivers and Streams," EPA 822-B-00-002, July 2000. http://www.epa.gov/waterscience/criteria/nutrient/guidance/rivers/rivers-streams-full.pdf

EPA. 2003. Survey of States, Tribes and Territories Nutrient Standards. May 2003.

EPA. 2006. Nutrient Criteria presentation. http://www.epa.gov/nps/tribal/pdf/r8nutrientcriteria.pdf

EPA. 2007a. Water Quality Trading Toolkit for Permit Writers. EPA-833-R-07-004

EPA. 2007b., Chapter 4, Antidegradation-Water Quality Handbook. July, 2007. http://www.epa.gov/waterscience/standards/handbook/chapter04.html

EPA. 2008. State Adoption of Numeric Nutrient Standards. EPA-821-F-08-007. Environmental Protection Agency, Office of Water, Washington, D.C.

EPA. 2009a. Office of Water, Office of Science and Technology, Science Advisory Board Review Draft, Empirical Approaches for Nutrient Criteria Derivation, August 17, 2009

EPA. 2009b. Letter to Administrator Lisa Jackson, Subject: An Urgent Call to Action – Report of the State-EPA Nutrients Innovations Task Group.

EPA. 2009c. An Urgent Call to Action: Report of the State-EPA Nutrient Innovations Task Group. August 2009.

EPA. 2009d. EPA letter from Ben Grumbles, Assistant Administrator, to Michael Sole, Secretary, Florida Department of Environmental Protection, January 14, 2009

EPA. 2009e. Environmental Appeals Board. Petition for Review, City of Twin Falls Wastewater Treatment Plan Permit No. ID-002127-0.

EPA, Office of Inspector General, "Evaluation Report, EPA Needs to Accelerate Adoption of Numeric Nutrient Water Quality Standards," Report No. 09-P-0223, August 26, 2009

http://www.epa.gov/oig/reports/2009/20090826-09-P-0223.pdf

EPA. 2010a. Science Advisory Board (SAB) Ecological Processes and Effects Committee Advisory Report, January 8, 2010.

EPA. 2010b. EPA Water Quality Standards Academy, Basic Course: Key Concepts, Variance: When Allowed. UALity
http://epa.gov/waterscience/standards/academy/keyconcepts/mod5/page11.htm

EPA. 2010c. Water Quality Standards, Module 15 Variances, 11-5-07.
http://www.epa.gov/waterscience/standards/academy/basic_course/15-variances-11-15-07.pdf

EPA. 2010d. Office of Water, Office of Science and Technology, Science Advisory Board Review, Empirical Approaches for Nutrient Criteria Derivation, April 27, 2010

Essig, D. and C. Grafe. 2005. Water Quality Standards. Idaho Department of Environmental Quality. Boise, ID.
http://www.deq.idaho.gov/water/assist_business/workshops/nps_workshop_05_essig_part2.pdf

FDEP. 2010a. Main Concerns with the Environmental Protection Agency's Proposed Numeric Nutrient Criteria for Florida's Lakes and Flowing Waters published January 26, 2010. Florida Department of Environmental Protection. February 17, 2010.
http://www.dep.state.fl.us/water/wqssp/nutrients/docs/federal/dep_concerns_epa_nutrients_0210.pdf

FDEP. 2010b. EPA's Proposed Numeric Nutrient Criteria for Lakes and Flowing Waters: Florida's Experience. January 26, 2010.
http://www.dep.state.fl.us/water/wqssp/nutrients/docs/federal/epa_nutrient_criteria_fdep_presentation_022310.pdf

Federal Register. 2010. Water Quality Standards for the State of Florida's Lakes and Flowing Waters; Proposed Rule. January 26, 2010.

Freedman, Paul, Len Shabman, and Kenneth Reckhow, Don't Debate; Adapt, Water Environment & Technology, Water Environment Federation, August 2008.

FSA. 2010. General Comments Concerning EPA's Proposed Numeric Nutrient Criteria. Florida Stormwater Association. http://ww.florida-stormwater.org/pdfs/2009-10%20PDFs/FSA_General_Comments_NNC.pdf

FWEA. 2010. Numeric Nutrient Criteria, Presentation to the Agricultural and Natural Resources Policy Committee. February 3, 2010. http://www.fl-counties.com/Docs/Legislative%20Division/Environmental/NNC%20Presentation%20to%20Fl%20House%20Committee%203feb09%20final%20(2).pdf

Hall. John C., and William T. Hall. 2009. Critical Evaluation of EPA Stream Nutrient Standard Initiatives. Environment Reporter, 40ER1593, Bureau of National Affairs, Inc. July 3, 2009.

Idaho Code. 2009. Title 39 Health and Safety Chapter 36 Water Quality, 39-3610 General Limitations on Point and Nonpoint Sources for Water Bodies Not Fully Supporting Beneficial Uses. Idaho Code, Boise, ID. http://www3.state.id.us/cgi-bin/newidst?sctid=390360010.K

IDAPA. 2009a. Idaho Administrative Code, Department of Environmental Quality, Water Quality Standards. Boise, ID. http://adm.idaho.gov/adminrules/rules/idapa58/0102.pdf

IDAPA. 2009b. Idaho Administrative Code, Department of Environmental Quality, Water Quality and Wastewater Treatment. Boise, ID. http://adm.idaho.gov/adminrules/rules/00codearchive/IDAPA16/0102.pdf

IDEQ. 1998. PM98-2 Policy for No-Net Increase. Idaho Department of Environmental Quality, Boise, ID. http://www.deq.state.id.us/about/policies/pm98_2.cfm

IDEQ. 2009. Lower Boise River Comments. Idaho Department of Environmental Quality, Boise, ID. http://www.deq.state.id.us/water/data_reports/surface_water/tmdls/boise_river_lower/boise_river_lower_comments_12to20.pdf

IGNN. 2006. Illinois EPA Grants Provisional Variance to City of Morrison. February 14, 2006

Illinois Environmental Protection Agency and The Village of New Lenox v. Illinois Pollution Control Board, Des Plaines River Watershed Alliance, Prairie Rivers Network, and Sierra Club. Appellate Court of Illinois. (3rd District, 2008).

Iowa, State of Iowa. Senate File 442 (SSB 1260). March 30, 2009.

Lee, G. Fred, and Anne Jones-Lee, PhD. 2008. Comments on US EPA's Conditional Probability Approach for Developing Phosphorus Nutrient Criteria. September 26, 2008.

Maryland (COMAR) Code of Maryland Regulation, http://www.dsd.state.md.us/comar/ (for 26.08.04.02; Accessed April 2, 2009).

Massachusetts Department of Environmental Protection. Antidegradation Review Procedure for Discharge Requiring a Permit Under 314 CMR 3.03. November 2008. http://www.mass.gov/dep/water/laws/antideg.doc

MDEQ. 2004. Wadeable Streams of Montana's Hi-line Region: An Analysis of Their Nature and Condition, with an Emphasis on Factors Affecting Aquatic Plant Communities and Recommendations to Prevent Nuisance Algae Conditions. Montana Department of Environmental Quality, Helena, MT.

MDEQ. 2008a. Scientific and Technical Basis of the Numeric Nutrient Criteria for Montana's Wadeable Streams and Rivers. Montana Department of Environmental Quality, Helena, MT.

MDEQ. 2008b. Circular DEQ-7 Montana Numeric Water Quality Standards. Montana Department of Environmental Quality, Helena, MT.

MDEQ. 2010a. Numeric Nutrient Criteria
http://deq.mt.gov/wqinfo/Standards/NumericNutrientCriteria.mcpx

MDEQ. 2010b. Administrative Rules of the State of Montana, Environmental Quality Chapter 30, Water Quality Subchapter 6. Helena, MT. http://www.deq.mt.gov/dir/Legal/Chapters/CH30-06.pdf

MEA. 2009a. Clean Water Advocates Announce Intent to Sue EPA to Set Standards Needed to Control Phosphorus and Nitrogen Pollution: Decade of Inaction Has Left Wisconsin Waters Unsafe. November 23, 2009 Press Release.

MEA. 2009b. RE: Notice of Intent to Sue Administrator of the Environmental Protection Agency for Failure to Perform Its Nondiscretionary Duty to Promulgate Numeric Nutrient Criteria for the State of Wisconsin. November 23, 2009. Midwest Environmental Advocates.

Missouri Department of Natural Resources. Nutrient Criteria Plan, Water Protection Program. September 13, 2005.

Missouri Department of Natural Resources. Missouri Antidegradation Rule and Implementation Procudure. April 20, 2007. Revised May 7, 2008. http://www.dnr.mo.gov/env/wpp/docs/aip-cwc-appr-050708.pdf

Montana (MCA) Montana Code Annotated, http://data.opi.state.mt.us/bills/mca_toc/75_5_3.htm (for 75-5-303. Nondegradation Policy).

New Mexico Environmental Department. Surface Water Quality Bureau. Total Maximum Daily Load (TMDL) for the Rio Hondo. June 14, 2005.
http://www.nmenv.state.nm.us/swqb/Projects/RioHondo/RioHondoTMDL2.pdf

North Carolina Department of Environment and Natural Resources. Division of Water Quality. Surface Waters and Wetlands Standards. North Carolina Administrative Code 15A NCAC 02B .0100, .0200 & .0300. May 1, 2007. http://h2o.enr.state.nc.us/admin/rules/codes_statutes.htm

NMC. 2010. Inland Numeric Nutrient Criteria: Ramifications for Tampa Bay's RA/TMDL. www.tbeptech.org/.../071_Numeric_Nutrient_Criteria_Ramifications-01222010_NMC_comments.ppt

NYU. 2010. Clean Water Act Summary.
http://74.125.95.132/search?q=cache:eckCqDAk3DUJ:www.law.nyu.edu/idcplg%3FIdcService%3DGET_FILE%26dDocName%3DECM_DLV_012385%26RevisionSelectionMethod%3DLatestReleased+A+water+quality+standard+variance+may+not+be+granted+if+other+dischargers+in+the+watershed+are+able+to+meet+the+water+quality+standards+and+effluent+limitations&cd=4&hl=en&ct=clnk&gl=us

TetraTech. 2008. Memorandum, Subject: Montana Stream Nutrient Criteria White Paper Review. Owings Mills, MD.

Tri-State Implementation Council. 1998. Nutrient Target Subcommittee, Clark Fork River Voluntary Nutrient Reduction Program (VNRP), August 1998. http://www.tristatecouncil.org/documents/98vnrp_doc.pdf

TSWQC. 2010. Clark Fork River Watershed Programs, Voluntary Nutrient Reduction Program. http://www.tristatecouncil.org/programs/cfrwatershed.html

UMRCC. 2008. Minutes of the Spring 2008 meeting of the UMRCC Water Quality Technical Section. http://www.mississippi-river.com/umrcc/Reports/Water%20Quality%20Page/WQ%20Tech%20Section%20Spring%202008%20Minutes.pdf

UMT. 2007. How the Public Views Nuisance Algae in Montana Streams. January 2007. http://www.umt.edu/watershedclinic/algaesurveypix.htm

United States Court of Appeals for the Ninth Circuit, Friends of Pinto Creek v. US EPA and Carlota Copper Company, Opinion No. 05-70785, October 4, 2007.

United States District Court for the Northern District of Florida, Florida wildlife Federation, Inc., Sierra Club, Inc., Conservancy of Southwest Florida, Inc., Environmental Confederation of Southwest Florida, Inc. and St. Johns Riverkeeper, Inc. V. Stephen L. Johnson, Administrator of the United States Environmental Protection Agency, Complaint for Declaratory and Injunctive Relief, 2008.

USC. 2010. Sec. 301 [33 USC 1311] Effluent Limitations http://www.usda.gov/rus/water/ees/pdf/cwa3.pdf

USGS. 2006. Nutrient Concentrations and Their Relations to the Biotic Integrity of Wadeable Streams in Wisconsin. Professional Paper 1722. USGS. Reston, VA.

USGS. 2008. Nutrient Concentrations and Their Relations to the Biotic Integrity of Nonwadeable Rivers in Wisconsin. Professional Paper 1754. USGS. Reston, VA.

USGS. 2001. Chemical and Biological Indicators of Nutrient Enrichment in the Yellowstone River Basin, Montana and Wyoming, August 2000: Study Design and Preliminary Results. Water-Resources Investigations Report 01-4238. Cheyenne, WY.

Water Environment Research Foundation. 2007. *Factors for Success in Developing Use Attainability Analyses*. Project # 04-WEM-1. Alexandria, VA.

Wyoming Department of Environmental Quality. Chapter 1, Water Quality Rules and Regulations-Wyoming Surface Water Quality Standards. February, 2007. http://soswy.state.wy.us/Rules/RULES/6547.pdf

CHAPTER 3.0

NUTRIENT CRITERIA ISSUES FOR DISCHARGES AND EFFLUENT LIMITATIONS

Nutrient driven water quality impairment is complex and loading conditions vary significantly from watershed to watershed across the country. Thresholds for nutrient enrichment vary, as do the magnitude of point and nonpoint source loadings, and waterbody responses. For these reasons, the potential water quality benefits from advanced wastewater treatment for nutrient removal will vary widely and depend largely on site specific circumstances in individual watersheds.

Wastewater utilities and private industry are dependent upon surface water discharges for effluent management and are subject to increasingly stringent nutrient limits. Even with alternatives to surface water discharge for reclaimed effluent reuse, seasonal weather limitations often necessitate surface water discharges when the need for reclaimed water is reduced or eliminated. Most wastewater treatment plants discharge to waterbodies that have been altered by other man caused activities and are not in a natural, undisturbed condition. Waterbodies that receive effluent discharges have impoundments and reservoirs, irrigation diversions and returns, water supply withdrawals, and many other modifications that alter the aquatic environment. These conditions present challenging circumstances for the selection of appropriate in-stream nutrient targets that protect water quality and reflect realistically attainable conditions.

This chapter describes some of the challenges associated with nutrient management in watersheds and requirements for wastewater discharge to surface waters. This includes a discussion of the issues associated with establishing appropriate in-stream nutrient limits for waterbodies, the importance of nonpoint source management, and legal issues that influence nutrient control considerations. Chapter 4.0 addresses advanced wastewater treatment technology for nutrient removal.

3.1 Dependency on Receiving Waters

Nutrients are transported to receiving water bodies by overland runoff of precipitation, groundwater, drainage networks, and waste effluents dischargers. Permitted dischargers are the most regulated source to receiving waters and, thus, must be aware of the permitting alternatives available.

If naturally occurring conditions in the receiving waterbody exceed nutrient criteria, then states may utilize one of two options to address the exceedance of the natural condition by either changing or removing the designated use or adjusting the existing criteria.

When naturally occurring pollutant concentrations prevent the attainment of a designated use, states may remove a designated use that is not an existing use, provided the state demonstrates the designated use is not attainable. States can also change the designated use by establishing subcategories of a use. A use attainability analysis (UAA) must be performed to change or remove a designated use.

An alternative to changing the waterbody's designated use is for States to establish site-specific numeric aquatic life water quality criteria by setting the criteria value equal to the natural background.

If a treatment technology does not exist to enable dischargers to meet a water quality based effluent limit (WQBEL), then states have the option of changing the water quality standards through variances or changes to designated uses. In such a case a new WQBEL would be calculated to meet the new use or variance. In other instances, dischargers may be able to meet the WQBELs based on existing water quality standards through options such as water quality offsets from point and nonpoint sources (e.g., land based BMPs), water quality trading, and watershed analysis. If dischargers cannot meet the nutrient criteria, they may pursue the process of changing the waterbody's designated use.

Another concern for dischargers is that they can be required to monitor constituents at points other than the effluent discharge outfall. Biological sampling may be appropriate to effectively monitor the discharge status and ensure compliance. One practice for collecting ambient monitoring is described in EPA's *Interim Guidance for Performance-based Reductions of NPDES Permit Monitoring Frequencies* (EPA, 1996), which states that the permit authority can grant reductions in effluent monitoring for a discharger with a history of good compliance and permitting performance in exchange for ambient monitoring.

Several pilot programs have been initiated to achieve balance between ambient monitoring and end-of-pipe monitoring. For example, Kodak Colorado Division (a division of Eastman Kodak) and other dischargers near Kodak on the Cache la Poudre River in northeastern Colorado have formed an ambient water quality monitoring group. The group was formed in cooperation with the Colorado Department of Public Health & Environment (CDPHE) to monitor the ambient water quality of the receiving waterbody (Kodak, 2006).

If a waterbody is currently listed as impaired, new sources and new dischargers still can be authorized. If a TMDL has been developed, the permit writer must demonstrate that there are remaining pollutant load allocations to allow for the additional loads and compliance schedules designed to bring the impaired waterbody into compliance with applicable water quality standards. When a TMDL has yet to be developed, the new source or new discharger can obtain a permit when certain conditions are met such as when the dischargers do not contain the pollutant causing the impairment, or other pollutant source reductions will offset the new discharge (EPA, 2008a).

3.2 Applicability of Ecoregion Criteria to "Real World" Receiving Waters

States have shown that many different methods can be used to determine and implement the EPA ecoregion nutrient criteria. The EPA (2000) cites five case studies in the *Nutrient Criteria Technical Guidance Manual for Rivers and Streams.* These examples are meant to illustrate real-world examples of nutrient criteria that can be developed on a practical level and several region-specific issues that may be encountered during the criteria development process.

The first case describes the Tennessee River system, where the Level III ecoregions were refined through the identification and monitoring of reference stream systems, and correlational analyses of nutrient levels, conventional water chemistry parameters, and biological indices were employed to derive criteria.

The second example is located in Clark Fork, Montana, where the state has delineated a process for setting target nutrient and algal levels based on a combination of modified established criteria, literature values, and observed thresholds for nuisance algal growth.

The Upper Midwest river systems case study describes the results of a cooperative effort among three USGS NAWQA projects in the upper Midwest Corn Belt region that evaluated algal and macroinvertebrate response to nonpoint agricultural sources relative to naturally-occurring factors (e.g., riparian vegetation, hydrology).

The Bow River, Alberta, Canada, case study details the reduction of nuisance biomass (both periphyton and macrophytes) over a 16-year period through decreases in nitrogen (approximately 50%) and phosphorus (80%) from domestic wastewater effluent.

Finally, the southwestern desert stream case study discusses several of the determinants of nutrient regimes in desert streams that should be considered when developing nutrient criteria for these, as well as other, complex, highly variable stream systems. (EPA, 2000).

Ecoregion criteria have not been met without criticism. Many of the concerns raised by the public about the EPA's approach for developing nutrient criteria were previously raised during the development of the EPA's technical guidance manuals.

The EPA's Nutrient Criteria Program later adopted the reference condition approach, and continues to recommend it in all of its nutrient criteria guidance manuals. Additionally, the statistical derivation approach to developing nutrient criteria was favorably reviewed. Consequently, the EPA did not change its fundamental approach to nutrient criteria development, or change the documents significantly beyond responding to comments of peer reviewers. Table shows a summary of scientific public comments received by the EPA (EPA, 2008).

Table 3-1. Summary of Significant Scientific Information Regarding the Development and Implementation of Ecoregion Criteria (EPA, 2008).

Topic	Description
Percentile Approach	Criteria based on statistics from regional data with gaps rather than site-specific data.
Site Specific Criteria	Aggregation of diverse conditions will not adequately protect a region's waters
Total Nitrogen Criteria	Total nitrogen criteria are not necessary.
Relation to Beneficial Uses	The EPA has not linked the criteria to beneficial uses.
CWA Requirements	Criteria do not meet requirements of Section 304 of CWA, which requires the EPA to develop water quality criteria that accurately reflect latest scientific knowledge.
Additional Work for States	States may be forced to spend additional time and effort defending criteria developed at state level that are different from the EPA ecoregion values.
Adverse Effects of Criteria	Criteria will harm the uses they are supposed to protect.
Effluent Dominated Streams	Ecoregion approach is based on assumption that waterbodies can achieve a natural, or "reference-reach" state, but EDWs cannot achieve such condition due to their hydrological and structural modifications, usually for drainage and flood control purposes.
Criteria are Not scientifically Valid	State efforts would be better directed toward monitoring and data collection, including an assessment of eutrophic conditions.
Criteria Have Adverse Impact with No Benefit	Because 304(a) nutrient criteria are not feasible, the foundation for evaluating attainment of water quality standards, setting targets for TMDLs, and establishing permit limits will be fundamentally flawed.

In the Pacific Northwest, construction of dams and the exclusion of migratory salmon from the upper reaches of streams has resulted in a significant decrease in nutrient fluxes from historical natural conditions. Ambient nutrient levels in some waters are now too low to support native and endangered fish species. To the extent that ecoregional reference data are based on a recent survey of nitrogen and phosphorus, they may be significantly lower than necessary to support native resident and anadromous fish populations.

Multiple state fisheries agencies (e.g. Idaho, Oregon, and British Columbia) in the Xeric West ecoregion add nutrients to streams to bring nitrogen and phosphorus concentrations up to a level that can support primary production and assist in endangered species recovery. Some examples include the phosphorus addition to Alturus Lake, Idaho, the Idaho Fish and Game nutrient addition to the Kootenia River, the Oregon Fish and Game salmon carcass distribution program, and the use of struvite (Ostara CrystalGreen) recovered from Clean Water Services of Washington County, Oregon Durham Wastewater Treatment Plant in salmon restoration programs in British Columbia.

3.2.1 Challenges with Irrigation Diversions, Dewatering, Reservoirs and Impoundments, and Artificial Flow Regimes

Waterbodies in economically developed areas used as receiving waters for effluent discharge are seldom found in their natural, undisturbed condition. Most have been altered over a long period of time by many manmade activities, including dams and impoundments, hydro-power facilities, water supply and irrigation diversions, agricultural return flows, artificially controlled flow regimes, and nonpoint source nutrient loadings. This raises the question of whether or not it is reasonable or feasible to expect that controlling point source nutrient discharges alone can meet water quality standards and attain beneficial uses based on natural conditions. While ecoregion nutrient criteria, macroinvertebrate indices, and other references for natural conditions provide useful guides for evaluating existing water quality, these may not be realistic endpoints for the altered waterbodies that are most often the effluent receiving waters. Beyond the pristine streams high in the watershed, many of our rivers have been managed and manipulated in ways that prevent the attainment of very low nutrient concentrations.

In his paper titled "Water Temperature and Dissolved Oxygen: Are State Water Quality Standards Achievable for Forest Streams?" George Ice wrote:

> "*...it is becoming increasingly evident that the TMDL process is overwhelming state environmental agency resources, resulting in TMDLs that are sometimes only slightly more than paper exercises. A widely recognized but little discussed problem with current state water quality standards (criteria) is that they are often unachievable even for the least-impaired watersheds. This means that environmental protection resources are not always being focused on real water quality problems and are being spread across both healthy and impaired watersheds. Unless achievable and biologically relevant criteria are identified it is likely that state agencies and private operators will be wasting resources on water quality goals that can never and should never be reached*" (Ice, 2006).

A study by Ice and Binkley revealed that:

"A review of 300 streams draining small forested watersheds finds that nutrient concentrations often exceed EPA's proposed criteria. More detailed nutrient concentration patterns were evaluated for eight unmanaged research forested watersheds, three of which would have failed EPA criteria and been identified as impaired" (Ice and Binkley, 2003).

If the headwaters of a stream originating in unmanaged forests fail to meet ecoregion criteria, how can we expect rivers that have been managed and modified to meet ecoregion criteria? In the Western Coalition of Arid States (WESTCAS) review of the EPA ecoregion nutrient criteria for the Xeric West, the criteria were found to be overly stringent and nearly every stream would exceed the criteria. Included in the WESTCAS review comments was the statement:

"Many western rivers are now dammed, and water is transported hundreds of miles through pipelines and canals. The changes have transformed rivers into reservoir, and parched desert into residential lawns. In many areas, runoff from lawn watering has created a perennial flow in what was formerly a dry wash. Due to these large-scale human modifications, it is not appropriate to base impairment on a comparison to "natural" waterways" (WESTCAS, 2001).

Eecoregion nutrient criteria are the EPA's recommendations for targeted in-stream water quality conditions, but they are not Clean Water Act requirements, nor are they state standards until adopted by a state through a rulemaking process. Unfortunately, these values can be readily applied to any waterbody and may not characterize the entire situation within a watershed in a balanced and practical way that reflects realistic expectations. Even in states without numerical standards for nutrients, EPA or the state may default to ecoregion reference points to support phosphorus and nitrogen limits. For example, the City of Leominster, Massachusetts argued that setting phosphorus limits in their NPDES permit was arbitrary and capricious. The EPA's response was that they simply used the ecoregion nutrient criteria as a means to interpret the state's narrative criteria (EPA, 2006).

3.3 Capabilities of Treatment Technology

Table 8 presents a simplified example of the *potential* capabilities of wastewater treatment technology at various levels. Influent concentrations of nitrogen and phosphorus in municipal wastewater are compared with a variety of potential effluent levels and with a range of in-stream nitrogen and phosphorus reference condition concentrations typical of ecoregion criteria in the Western US (Eco-region I. Willamette and Central Valley, II Western Forested Mountains, III Xeric West, and IV Great Plains). Secondary treatment facilities produce effluent nutrient concentrations approximately the same as influent wastewater, with limited removal due to synthesis in biological treatment.

The entry level of nutrient removal facilities, designated in Table 3-2 as biological nutrient removal (BNR) is typically capable of producing effluent phosphorus of approximately 1 mg/l and effluent nitrogen of approximately 10 mg/l. Enhanced nutrient removal, or ENR, with effluent filtration and larger biological reactors, may produce effluent phosphorus in the range of 0.25 to 0.5 mg/l and effluent nitrogen of 4 to 6 mg/l. The most advanced nutrient removal systems operating at the maximum capability of treatment technology with multiple filtration steps or membranes, and larger biological reactors, may reduce effluent phosphorus to approximately 0.05 to 0.07 mg/l and effluent nitrogen to 3 to 4 mg/l.

Comparing the very best nutrient removal facilities with potential numeric nutrient targets based on ecoregion reference criteria reveals that effluent phosphorus may approach these in-stream targets should they be applied end-of-pipe. However, even the very best nitrogen removal facilities would have effluent much higher than in-stream target levels typical of the Western US. Since in-stream numeric nutrient criteria based on natural conditions are very low concentrations, they may result in very restrictive discharge permit limits that are lower concentrations than treatment technologies are capable of achieving if applied "end-of-pipe."

This is a concern since wastewater utilities rely on surface waters for effluent management. Most discharges are to streams that are altered in many ways resulting in conditions far from natural. Overly restrictive effluent limits for Point Sources may have unintended consequences that are not beneficial, such as diversion of effluent from streams with limited flow. Further, reduction in point sources alone will not improve water quality since nonpoint sources make up the majority of nutrient loadings in most waters.

Table 3-2. Generalized Comparison of Phosphorus and Nitrogen Concentrations for Wastewater, Effluent from Advanced Treatment[a] and Typical In-Stream Nutrient Criteria[b]

Parameter	Typical Municipal Raw Wastewater, mg/l	Secondary Effluent (No Active Nutrient Removal), mg/l	Typical Advanced Treatment Nutrient Removal (BNR), mg/l	Enhanced Nutrient Removal (ENR), mg/l	Limits of Treatment Technology, mg/l	Typical Instream Nutrient Criteria[b], mg/l
Total Phosphorus	4 to 8	4 to 6	~1	0.25 to 0.50	0.05 to 0.07	0.010 to 0.050
Total Nitrogen	25 to 35	20 to 30	~10	4 to 6	3 to 4	0.1 to 0.600

[a] The expected nutrient treatment removal levels and associated effluent concentrations vary widely according to the averaging period and/or performance statistic employed.

[b] Typical in-stream targets generalized from Eco-region I. Willamette and Central Valley, II Western Forested Mountains, III Xeric West, and IV Great Plains – see Table 2-2.

Chapter 4.0 presents a more detailed discussion of advanced wastewater treatment technology for nitrogen and phosphorus removal. The current state-of-the-art is summarized and the limits of treatment capability are described. At low effluent limits, some portion of the remaining nitrogen and phosphorus in treatment plant discharges may not be removable with current treatment technology. Nitrogen and phosphorus speciation is an important area of nutrient research, both in terms of biodegradability in wastewater treatment and bioavailability in the water environment.

Typical removal capabilities with industrial wastewater is more difficult to characterize due to widely varying compositions and nutrient levels. A significant percentage of industrial wastewater treatment systems are not capable of reaching the same "limits of technology" as municipal systems due to either much higher influent levels, or refractory species of nitrogen and phosphorus as discussed below.

3.3.1 Phosphorus Speciation

Phosphorus speciation refers to the different forms of this nutrient that exist in a waterbody. Nutrients can be categorized many ways. When considering TMDLs for rivers, lakes, and reservoirs, the common forms of phosphorus assessed include: inorganic (such as orthophosphate) and organic phosphorus.

Phosphorus speciation has become an increasingly important consideration in the management of nutrient loadings in sensitive watersheds. This includes watersheds such as the Spokane River with a very low in-stream target concentration of 10 μg/L total phosphorus. Nonpoint source dominated watersheds such as the Florida Everglades also face similar challenges of very low phosphorus concentration targets. Phosphorus speciation (total, particulate, soluble reactive or ortho-phosphate, and soluble organic or non-reactive) may be indicative of the sources of loadings within the watershed and aid in nonpoint source loading analysis and source tracing. Phosphorus speciation also provides an indication of how bioavailable the nutrients loads may be to drive the enrichment that leads to dissolved oxygen depression.

Phosphorus speciation may be a key to the potential for removal in wastewater treatment facilities and in nonpoint source best management practices. Advanced nutrient treatment processes are limited in their ability to remove soluble, **r**efractory **d**issolved **o**rganic **p**hosphorus (RDOP). Residual levels of RDOP in the effluent from the most advanced wastewater treatment processes are on the same order as the in-stream total phosphorus targets.

3.3.2 Nitrogen Speciation

The importance of nitrogen speciation has also become apparent in key watersheds of concern, such as Chesapeake Bay. Current scientific investigations are underway to attempt to understand the potential bioavailability of dissolved organic nitrogen in the aquatic environment. For example, one study is looking at the effects of exposure to salinity and sunlight which could result in dissolved organic nitrogen becoming available as a nutrient source with time in the marine environment.

For point source dischargers, effective treatment processes are available for removal of inorganic nitrogen (ammonia, nitrate, nitrite). However, advanced nutrient treatment processes are limited in their ability to remove soluble, **r**efractory **d**issolved **o**rganic **n**itrogen (RDON). Residual levels of RDON in the effluent from the most advanced wastewater treatment processes can exceed levels of in-stream total nitrogen targets by several times.

3.3.3 Treatment Technologies

Bott, Murthy, et al. (2006) conducted a comprehensive state-of-the-science workshop, including about 120 world-renowned subject matter experts and stakeholders in Washington, D.C. in 2006. This three-day workshop brought together for the first time designers, practitioners, regulators, operators, and researchers to share their diverse perspectives, experiences and expertise on removing nutrients, as well as to discuss the limits of technology (LOT) that currently prevent further reduction.

Participants addressed topics such as "boundary conditions" (factors that limit nutrient removal and treatment processes) including why and when meeting certain low levels may not be feasible. Boundary conditions include low temperature; period of high flow (wet weather); recycle, centrate, and side-stream flows; refractory dissolved organic nitrogen; and inhibition of nitrogen and phosphorus removing organisms or mechanisms. Participants also shared experience on design, operations, and modeling. Data and case studies from various facilities, innovative treatment processes, and regulatory approaches were discussed.

It was found that there are currently no universal answers or approaches for wastewater nutrient removal. While there are some successful processes for nitrogen or phosphorus removal, there are fewer for the simultaneous removal of both N and P at low levels (<3 mg/L for N, <0.1 mg/L for P) in a consistent and cost-effective manner. It was also determined that engaging regulators and stakeholders are essential to solving the problem, that LOT and boundary conditions need to be further defined, and that additional research is needed in several areas including the speciation and characterization of both N and P.

The report, summarized in 05CTS1W (2006) includes a CD-ROM of the PowerPoint presentations from this workshop, including an introduction to the subject by experts, case studies, and data from advanced wastewater treatment facilities both in the U.S. and abroad. The report identified and prioritized short, medium, and long-range research and actions needed to address this complex issue. Knowledge gained from this workshop serves as the basis and roadmap for WERF's ongoing multi-year research in this area, the Nutrient Removal Challenge.

Neethling, et al. (2005) surveyed a number of municipal wastewater plants from various countries and found that EBPR (enhanced biological phosphorus removal) is capable of achieving very low effluent phosphate concentrations. When operating well, phosphate concentrations <0.1 mg/L could potentially be achieved for extended periods (more than a month), 0.03 mg/L for a week, and even below 0.02 mg/L for several sequential days. However, excursions above these levels were common. EBPR performance can be estimated by examining plant influent BOD and total P loading, the level of oxidants in the anaerobic zone, the degree of recycle phosphorus control, and operating parameters (solids retention time, temperature, dissolved oxygen, anaerobic and aerobic hydraulic retention time).

Strom, et al. (2005) developed the first application of CFD (computational fluid dynamics) to activated sludge biological treatment systems to clarify the role of the bioreactor macro-environment in SBNR (simultaneous biological nutrient removal), defined in their study as the removal of nitrogen or phosphorus in excess of that required for biomass synthesis in biological wastewater treatment systems where defined anaerobic or anoxic zones do not exist. The CFD model simulated the creation of dissolved oxygen gradients within the system, demonstrating that the anaerobic zones required for SBNR could occur. This work is an important step towards the development of a mathematical model for realistic closed loop bioreactors and has the potential to enhance nutrient removal economically and more reliably in existing facilities.

Some advanced wastewater treatment processes have been shown to be effective in reducing N or P to fairly low levels; however, it is currently not possible to remove both N and P simultaneously in a sustainable and cost-effective manner. There are limits to technology (LOT) and site-specific conditions that limit their effectiveness. Some of these processes are considered tertiary (not secondary) treatment technologies.

3.3.4 Ongoing Efforts and Additional Resources

WERF's Nutrient Removal Challenge is a multi-year effort focused on helping wastewater utilities achieve the most efficient and cost-effective nutrient removal technically feasible in order to meet permit limits and sustain treatment operations. It seeks to better understand existing mechanisms of nutrient removal, best available technologies, and limits of technologies so regulators and permittees can make more informed decisions. An online information exchange system was created to share expert knowledge in a collaborative and timely manner (WERF, 2010).

Two pre-conference workshops were held on this subject at WEFTEC 2008 in October. The first (W201) will provide an update on activities and findings from the EPA, WERF, and other partners, and the second (W210) will be on the LOT and statistical analysis of long-term operating data from over a dozen nutrient removal facilities nationwide (WEF/WERF, 2008).

In addition, WERF hosted the third in a series of annual web seminars on this subject in late 2008. Previous seminars have been recorded and archived. Details are available on the WERF website (WERF, 2010).

The EPA is currently completing a technology resource document for use by state agencies and permit writers. The two volume document utilizes information from Pagilla, et al. (02CTS1, 2008), including technology screening criteria/protocols as well as data from plants participating in the WERF study (Pagilla, 2008). This work was presented during the WEF/WERF preconference workshop W201 at WEFTEC 2008 in Chicago (WEF/WERF, 2008).

The EPA Office of Wastewater Management's Municipal Technologies website has other resources available online (EPA, 2008d). Links to these and other resources are available from the WERF nutrient removal challenge web portal (WERF, 2010).

3.4 Point Source v. Nonpoint Source Issues in Watershed Management

Point and nonpoint source pollution are the two most general categories of water pollution. Point sources originate at a single location or outlet and are easy to measure. Nonpoint source (NPS) pollution, unlike pollution from industrial and sewage treatment plants, comes from many diffuse sources and is more difficult to detect and quantify in many cases. NPS pollution is caused by rainfall or snowmelt moving over and through the ground. As the runoff moves, it picks up and carries away natural and human-made pollutants, finally depositing them into lakes, rivers, wetlands, coastal waters, and even in underground sources of drinking water. Agricultural runoff is the largest source of nonpoint source nutrient loading, primarily due to fertilizer application to cropland.

According to EPA, most watersheds are impaired by a combination of point sources and nonpoint sources. In some watersheds, impairment is dominated by nonpoint sources. In watersheds where nonpoint source nutrient loadings outweigh point sources, advanced treatment for nutrient removal and even complete elimination of point sources by zero discharge may have limited effect on water quality.

In most waterbodies, point source wastewater discharges are only a part of the total nutrient loading to the watershed. According to EPA, most watersheds are impaired by a combination of point sources and nonpoint sources. Impairment in most watersheds is caused by a combination of point and nonpoint sources, or is dominated by nonpoint sources. Without nonpoint nutrient controls, technology based nutrient standards for wastewater discharges would have limited benefit for waterbodies nationally.

In watersheds where nonpoint source nutrient loadings outweigh point sources, advanced treatment for nutrient removal and even complete elimination of point sources by zero discharge may have limited effect on water quality. Nutrient loading summaries for the Gulf of Mexico, Chesapeake Bay, and Flathead Lake, Montana are presented in Figure 3-1 for phosphorus and Figure -2 for nitrogen. Point source phosphorus loadings in these three key watersheds range from as little as 2% in the Flathead Lake watershed, to 22% in Chesapeake Bay, and 34% in the Gulf of Mexico. Point source nitrogen loadings range from as little as 2% in the Flathead Lake watershed, to 20% in Chesapeake Bay, and 22% in the Gulf of Mexico.

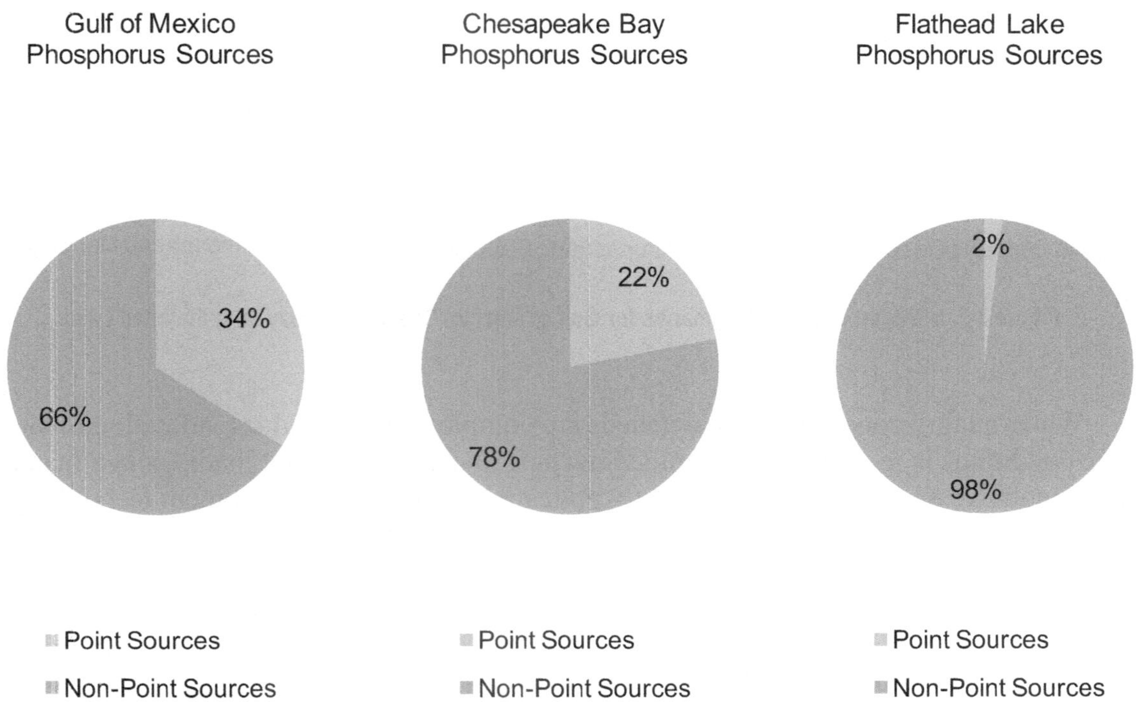

Figure 3-1. Phosphorus Loading Summaries for Gulf of Mexico, Chesapeake Bay, and Flathead Lake.

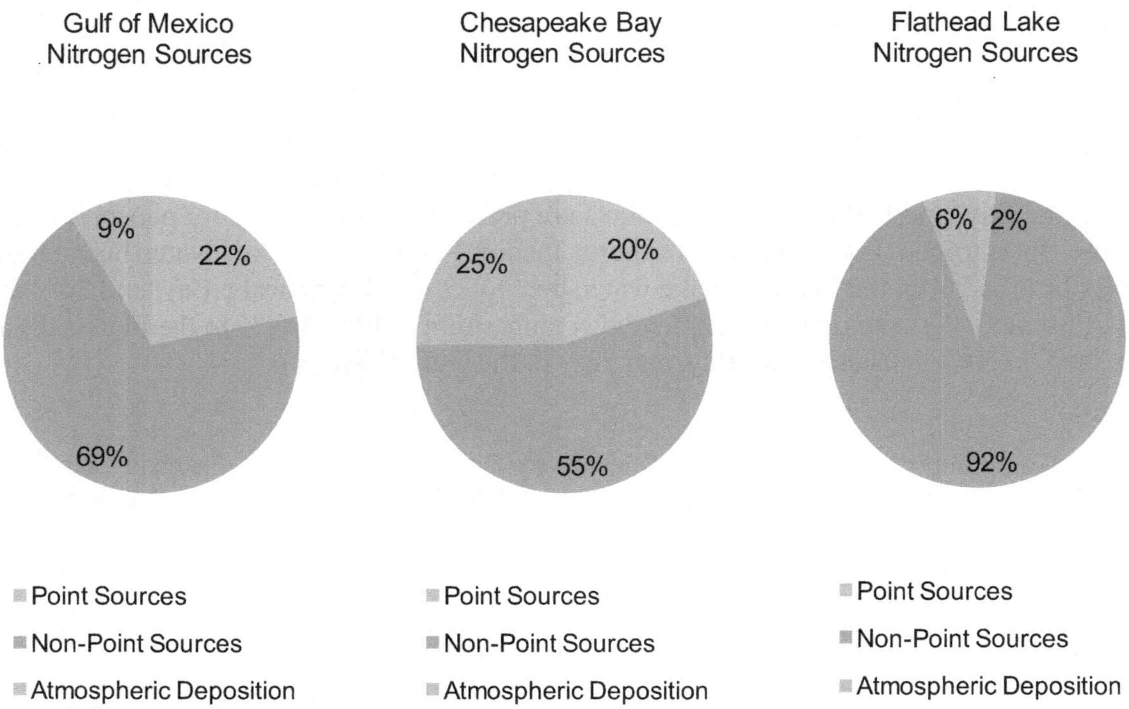

Figure 3-2. Nitrogen Loading Summaries for Gulf of Mexico, Chesapeake Bay, and Flathead Lake.

Water quality conditions in watersheds are complex and careful consideration of site specific conditions is required to determine how best to manage nutrient loadings in a manner that is economical and technically feasible. Application of wastewater treatment technology alone will have limited potential to provide water quality benefits if nonpoint sources are not controlled. However, nonpoint source control of phosphorus and nitrogen loadings has great potential for substantial reduction of nutrient loadings to improve water quality.

3.4.1 Nonpoint Source Pollution

Nonpoint source management practices may represent the best way to achieve overall environmental benefits in a watershed for a number of reasons. Many nonpoint source controls can address nutrient impairments more cost effectively than point source reductions, especially as wastewater treatment requirements approach the limits of technology. Further, nonpoint source management practices can achieve suspended sediment reductions which point sources cannot. Siltation, sedimentation, and bacterial impairments tend to occur concurrently with nutrient impairment and nonpoint source controls can address multiple water quality impairments simultaneously. Nonpoint source controls can also improve habitat quality for biological resources by reducing peak stream flow velocity (scouring) and providing greater riparian shading of streams (reduces temperature, improves dissolved oxygen). Nonpoint source controls can sequester carbon from the watershed (e.g. riparian buffers assimilate carbon) and use less energy than wastewater treatment plants, which reduces greenhouse gas emissions.

3.4.2 USGS Scientific Investigations Report

The USGS examined three sources nutrients to the land surface, nitrogen and phosphorus from fertilizer (farm and nonfarm use), livestock manure, and nitrogen from atmospheric deposition (USGS, 2006). These nonpoint sources of nutrients are typically readily available and mobile leading to the transport of these nutrients in stormwater runoff and groundwater that eventually contribute to surface waters.

A variety of data sources were used to examine the available nitrogen and phosphorus data for fertilizer, manure, and atmospheric deposition. These data included fertilizer sales, expenditures on fertilizer, livestock populations, human populations, and wet deposition chemistry. The USGS used these data to estimate the temporal and spatial patterns for nutrient loads to the land surface.

The USGS found that the largest source of nitrogen to the land is farm fertilizers at approximately 56% of the total from all sources. The total annual loading of nitrogen to the land surface was estimated at nearly 20 billion kilograms in 1997. This is an increase in farm fertilizer nitrogen from about 1 billion kilograms in 1950 to about 11 billion kilograms in 1997 (USGS, 2006).

The USGS found that the largest sources of phosphorus to the land are farm fertilizer and livestock manure, at approximately 50 and 49% of the total from all sources, respectively. The total annual loading of phosphorus to the land surface was estimated at nearly 3.5 billion kilograms in 1997. This is an increase in farm fertilizer phosphorus from about 0.6 billion kilograms in 1950 to about 1.8 billion kilograms in 1997 (USGS, 2006).

The USGS was only able to examine nitrogen on a spatial scale. They found that nitrogen from farm fertilizers were highest in the upper Midwest, along the east coast, extensive agricultural areas from Nebraska to Texas, and the intensively irrigated areas of southern Idaho, eastern Washington, northwestern Oregon, and central California. Nitrogen from livestock manure was highest in the South-Central and Southeastern U.S. and central California. These areas with the highest farm fertilizer had application values greater than 8,000 kilograms per square kilometer. Nitrogen from non-farm fertilizer was highest around most metropolitan areas although the highest values were greater than 2,000 kilograms per square kilometer. Nitrogen from atmospheric deposition was highest in the Central and Eastern U.S. although the highest values were greater than 500 kilograms per square kilometer (USGS, 2006).

Farm fertilizers and livestock manure quantities continue to increase. Fertilizers and manure are potential nonpoint sources of nitrogen and phosphorus requiring best management practices. Without proper application, use, and control, these nutrients can be transported to surface waters and impair water quality.

3.5 Significant Costs of Nutrient Removal

Nutrient removal requires additional treatment facilities beyond secondary treatment, which results in increases in both capital and operating costs. Costs for treatment plant retrofits for nutrient removal are highly dependent upon existing infrastructure and the extent to which facilities modifications are required to meet effluent nitrogen and phosphorus limits. For these reasons, it is difficult to generalize about the costs for modifications required for nutrient removal in a manner applicable across the country and costs vary widely for both retrofit facilities and new plants.

Nutrient removal also requires additional energy, chemicals, maintenance materials, and labor which increase operations and maintenance costs. When chemicals are added for nutrient removal, additional solids must be processed in the treatment plant and managed in biosolids utilization or disposal programs. Increases in solids loadings also increase capital, operating, and disposal costs for wastewater treatment.

The EPA recently published a 12-page fact sheet on biological nutrient removal (BNR) technologies, removal efficiencies, and associated costs (capital and operating) for small and large municipal systems (EPA, 2007).

EPA has also recently published a Municipal Nutrient Removal Technologies Reference Document evaluating the performance and costs of facilities removing nitrogen and phosphorus (EPA, 2008). This provides a broad survey of relatively recent nutrient removal costs at various effluent levels. EPA examined effluent nitrogen and phosphorus performance at 29 full-scale treatment plants in the United States and one in Canada. Detailed process information and costs were analyzed for more than 40 different treatment technologies for removing nitrogen and phosphorus from municipal wastewater. Nine facilities were studied in depth with case studies presented in an appendix. The case studies used performance data from a one year period to identify the factors influencing performance, reliability, and costs.

3.6 Sustainability

Advanced wastewater treatment facilities, including plants that remove/reduce N and/or P require significant capital costs and operating expenses, including energy and chemicals, and complex operational expertise and staffing needs. There has been some concern raised about the sustainability of these operations, the workforce needed, and increases to sewer and water rates to the public. In addition, there are issues related to the potential contribution to greenhouse gas emissions from the NOx emissions, energy shortages, the rising costs of fuel based polymers and chemicals, and climate change issues. Chemicals, such as alum and ferric compounds needed in some processes to remove phosphorus to very low levels, increase the quantities and affect the quality of the treated biosolids that have to be managed. Phosphorus is also a very limited resource that can potentially be harvested (recovered) from wastewater. Nutrients typically in high quality treated biosolids are an excellent resource for farmers and gardeners. Studies and more holistic life cycle analyses are currently underway both within and outside WERF to better understand these complex issues and how to benefit the public.

3.6.1 EPA's Sustainability Goals

EPA's strategic plan also includes an emphasis on sustainability themes, including reduction of greenhouse gas emissions and sustainable agriculture. EPA is required to update its strategic plan every three years by the Government Performance and Results Act of 1993. The 2009-2014 EPA Strategic Plan Change Document highlights only those strategies and performance measures significantly different from the 2006-2011 strategic plan (EPA, 2008). The schedule for updates fell during the transition of administrations and as such the current administration may promote some of the goals more aggressively than others identified. Additionally EPA identified that they will need to "continue to refine the strategic measures and establish baselines and targets throughout the Strategic Plan update process" (EPA, 2010).

EPA grouped the new or significantly different strategies into five goals (EPA, 2008). Elements of these goals are wide reaching, touching almost every aspect of EPA's environmental regulation. These goals are:

- Goal 1 – Clean Air and Global Climate Change
- Goal 2 – Clean and Safe Water
- Goal 3 – Land Preservation and Restoration
- Goal 4 – Healthy Communities and Ecosystems
- Goal 5 – Compliance and Environmental Stewardship

Under Goal 1, the clean air and global climate change, greenhouse gas emissions may have the greatest impact on wastewater treatment systems. With regard to greenhouse gas emissions, EPA plans to expand its voluntary programs. The strategic measure is projected to reduce 185 million metric tons of carbon equivalents by 2014 (EPA, 2010). Under Goal 1, EPA will focus research on the identification of adaptation strategies that yield additional benefits ("co-benefits") in the form of reductions of greenhouse gas emissions. Among these "co-benefits" are smart growth policies that lead to reductions in emissions of criteria air pollutants and improvements in air quality. The agency's air program will examine policy and management approaches to adaptation and conduct analyses to estimate the economic implications of climate change and the benefits of adapting to climate change. Research-based decision-support tools will be developed for stakeholders in states and local communities to help implement adaptation strategies and incorporate climate change elements into their day-to-day operations. Targeted large atmospheric emission sources, such as agriculture, present an opportunity to reduce both nutrient loads and GHGs by offsetting fertilizer manufacture with recovery and reuse of fertilizer recovered from wastewater treatment.

With regard to treatment of wastewater, nutrient reductions, recovery, and reuse using biological processes that minimize greenhouse gases related to net energy and chemical use may be a challenge when confronted with very low concentrations of nitrogen and phosphorus limits. NPDES permitting strategies that incorporate GHG emissions as part of the target and permitting was not identified by EPA as a strategy, but needs to be addressed to balance competing environmental protection targets. Nutrient control requirements that call for treatment at, or beyond the capabilities of technology maybe be counterproductive in balancing overall environmental goals. Further, opportunities exist to couple advanced treatment for nutrient removal with nutrient recovery in a manner that captures multiple benefits.

Struvite recovery is an example of a process to reduce phosphorus and ammonia recycle loadings from wastewater facilities with anaerobic digestion and dewatering that reduces the cost of treatment, reduces chemical use, reduces GHG emissions, and generates a commercial fertilizer product. The production of a fertilizer product is aligned with EPA's strategic plans for sustainable agriculture and is also linked with the future worldwide limits in phosphate fertilizer. For the past century, agriculture has relied increasingly on mined phosphate rock to achieve high crop yields and the fertilizer industry acknowledges that remaining reserves are decreasing (Cordell, 2009). Phosphate rock resources are finite and demand for phosphorus fertilizers are expected to exceed supply by 2033 when production is projected to peak (Cordell, 2009).

EPA's clean and safe water Goal 2 has the greatest number of strategies that may impact utilities and private industry including climate change, contaminants of emerging concern, security, and environmental indicators and monitoring. One strategy is to begin implementing the National Water Program Strategy: Response to Climate Change. This means integrating the impacts of climate change on EPA's water programs. These impacts could mean addressing potential changes in precipitation and streamflow in analyses completed for TMDLs and NPDES permitting. EPA plans to address contaminants of emerging concern, such as pharmaceuticals, personal care products, and nanomaterials by using a four-pronged approach. Of these four approaches, the ones that will have the most impact are: "preventing their entry into our waterways and promoting good stewardship and taking regulatory actions where appropriate" (EPA, 2010). Continued updates to strategic measures include protecting water quality and improving water quality on a watershed basis. The measures include reducing impairments and attaining water quality standards for more waterbodies by 2014.

The healthy communities and ecosystems Goal 4 includes identifying specific geographic areas where EPA will pursue additional strategies for protection. Specific areas include the Great Lakes, Gulf of Mexico, the Upper Mississippi River Basin, and the Delta/San Francisco Bay Estuary. These areas along with the Chesapeake Bay Ecosystem, Long Island Sound, South Florida Ecosystem, Puget Sound Basin, and Columbia River Basin have specific and quantified strategic measures to meet by 2014. The Gulf of Mexico measure includes reducing the release of nutrients throughout the basin by 2015 to reduce the hypoxic zone.

Under the compliance and environmental stewardship Goal 5, is a strategy to revise the enforcement and compliance measurement approach. "EPA's enforcement and compliance program is restructuring its measurement system from a tool-based approach to a problem-based approach." "Problem-based performance reporting will move progressively toward characterizing pollutant loadings with the ultimate goal of providing data on ecological and human health benefits" (EPA, 2010). The strategic measures include addressing environmental problems from water pollution. The measure is "by 2014, reduce, treat, or eliminate (an un-quantified) estimated pounds of water pollutants." "EPA will break out the 'environmental significant' water pollutants that affect the top 5-10 causes of impairment to waters" (EPA, 2010). These pollutants include *nutrients*, pathogens, mercury, other metals, sediment/turbidity, toxic organics, pH, temperature, and salinity.

The initiatives the administration promotes and the actions EPA undertakes to meet these strategies and performance measures in next four to five years could have significant impacts on both the environment and utilities.

3.6.2 Sustainable Nutrient Management

Nutrient removal treatment facilities have a larger environmental impact than secondary treatment as a result of the additional energy consumption and greenhouse gas emissions (GHG) from advanced wastewater treatment processes. Carbon footprint is used to describe the greenhouse gas emissions associated with an activity such as wastewater treatment. Greenhouse gas emissions occur directly from the wastewater treatment plant processes and indirect emissions occur from the consumption of electrical power. Advanced wastewater treatment facilities for nutrient removal both emit more greenhouse gases directly and consume more electrical power.

Decisions to require advanced treatment for nutrient removal should be considered carefully in order to avoid unnecessarily increasing greenhouse gas emissions with marginal water quality benefits. Since advanced treatment will consume more electrical power and generate more greenhouse gas emissions, care should be taken to avoid unnecessarily restrictive effluent nutrient requirements that will adversely affect other parts of the environment and public health.

Table 3-3 presents a summary comparison of nutrient management approaches for point and nonpoint sources, including a comparison of nutrient removal performance and potential costs. Point source wastewater treatment performance in removing nutrients is predictable over a very specific range with detailed data available for analysis. Nonpoint source best management practice (BMP) effectiveness is more uncertain, with potential nutrient control performance varying over a wide range.

Costs for both point and nonpoint source controls vary widely, often depending upon the extent and condition of existing treatment facilities and the site specific land activities occurring in the watershed. The most cost effective nutrient reductions to make in a given watershed are dependent on site specific circumstances that are important to understanding the conditions in individual treatment facilities and watersheds.

Sustainability is an important consideration in nutrient management and Table 3-3 highlights contrasting impacts and benefits associated with point source and nonpoint source approaches. More advanced levels of point source wastewater treatment come with penalties in terms of additional energy use, chemical use, and both direct and indirect greenhouse gas emissions. Advanced treatment facilities provide little aesthetic benefit to watersheds. In contrast, nonpoint source management practices do not consume electrical energy or use chemicals, and instead of emitting greenhouse gases, they may sequester carbon. Further benefits of nonpoint source controls are enhanced watershed habitat and aesthetics.

This comparison illustrates the importance of a balanced consideration of point and nonpoint source controls for effective watershed management and water quality protection. Towards that end, WERF is undertaking a study of nutrient control and sustainability titled "Finding the Balance Between Wastewater Treatment Nutrient Removal and Sustainability, Considering Capital and Operating Costs, Energy, and Air and Water Quality" that is planned for publication in 2010.

Table 3-3. Summary Comparison of Point and Nonpoint Source Nutrient Management[a]

Management Approach	Performance, % Reduction		Cost Effectiveness, $/lb Removed		Electrical Power Consumption	Chemical Use	Greenhouse Gas Emissions	Watershed Enhancements (Habitat, Aesthetics, Sediment Reduction)
	Nitrogen	Phosphorus	Nitrogen	Phosphorus				
Point Source Nutrient Reduction								
Advanced Wastewater Treatment	80 – 90[a]	90 – 99	$0.50 - $3.30	$2.60 – $37.00	50% - 250% increase over Secondary Treatment	Alum, Ferric, Methanol, other carbon sources	120% increase over Secondary Treatment	None
Nonpoint Source Best Management Practices (BMPs)								
Conservation Tillage	15	66	$0.05 – $3.30	$2.60 – $150	None	None	Sequesters Carbon	Moderate
Grass Buffers	50 - 80	50 - 80	$0.60 – $17.00	$11.00 – $190	None	None	Sequesters Carbon	Moderate
Detention Basins	30 - 65	30 - 65	$110	$320 – $700	None	None	Sequesters Carbon	High
Wetlands	<30	15 - 45	$2.20	$43.00 – $52.00	None	None	Sequesters Carbon	High

[a] Reprinted with permission from NACWA. 2009. NACWA Nutrients Issue Paper, Technical Discussion

3.7 References

Dependency on Receiving Waters:

EPA. 1996. Interim Guidance for Performance-based Reductions of NPDES Permit Monitoring Frequencies.

EPA. 2008a. Nutrient Criteria Implementation, http://www.epa.gov/waterscience/criteria/nutrient/faq/, Accessed: September 6, 2008.

EPA. 2008b. www.epa.gov/owow/watershed/trading/mabe.ppt, Accessed: September 8, 2008.

Kodak. 2006. Environmental Facility Report for 2006, Kodak Colorado Division Eastman Kodak Company, Windsor, CO.

Applicability of Ecoregion Criteria to "Real World" Receiving Waters:

EPA. 2000. Nutrient Criteria Technical Guidance Manual: Rivers and Streams, Washington, D.C., EPA-822-B-00-002.

EPA. 2008. http://www.epa.gov/waterscience/criteria/nutrient/ecoregions/files/pubcom.pdf

EPA. 2006. NPDES Permit # MA0100617 2006 Reissuance Response to Public Comments, EPA, Washington, D.C.

Ice, G.G. 2006. Water Temperature and Dissolved Oxygen: Are State Water Quality Standards Achievable for Forest Streams? Proceedings of 9th Biennial Watershed Management Conference, Center for Water Resources Report No. 107, University of California, Riverside, CA.

Ice, G.G., and Binkley, D. 2003. Forest streamwater concentrations of nitrogen and phosphorus: A comparison with EPA's proposed water quality criteria. Journal of Forestry 101(1):21-28.

Western Coalition of Arid States (WESTCAS). 2001. Comments on Ambient Water Quality Criteria Recommendations Rivers and Streams in Nutrient Ecoregion III (Xeric West) EPA 822-B-00-016. Letter to Robert Cantilli, U.S. EPA, Washington, D.C.

Treatment Technology:

Bott, C.B., Murthy, S.N., Spano, T.T., Randall, C.W., 2006. Workshop on Nutrient Removal: How Low Can We Go & What is Stopping Us from Going Lower? Project # 05CTS1W. Water Environment Research Foundation. Alexandria, VA.

Neethling, J.B., et al, 2005. *Factors Influencing the Reliability of Enhanced Biological Phosphorus Removal*, Project # 01-CTS-3. Water Environment Research Foundation. Alexandria, VA.

Pagilla, K., et al, 2008. *Sustainable Technologies to Achieve Very Low Nitrogen and Phosphorus Effluent Levels*, Project #02-CTS-1. Water Environment Research Foundation. Alexandria, VA.

Strom, P.F, Daigger, G., Littleton, H.X. 2005 *Characterizing Mechanisms of Simultaneous Biological Nutrient Removal During Wastewater Treatment*, Project #00-CTS-17UR. Water Environment Research Foundation. Alexandria, VA.

WEF/WERF. 2008. "*Demonstrated Processes for Limit of Technology Nutrient Removal – Achievable Limits and Statistical Reliability*." C. Bott, D. Parker, JB Neethling (chair).

Preconference Workshop 101, WEFTEC 2008, Water Environment Federation Technical Exhibition and Conference, Chicago, IL.

WEF/WERF. 2009. Real World LOT Nutrient Removal: Achievable Limits and Statistical Reliability. C. Bott and D. Parker (chair). Preconference Workshop 216, WEFTEC 2009, Water Environment Federation Technical Exhibition and Conference, Orlando, FL.

WERF. 2010. Knowledge Area: Nutrients. http://www.werf.org/AM/Template.cfm?Section=Nutrients&Template=/TaggedPage/TaggedPageDisplay.cfm&TPLID=23&ContentID=5420

Point Source vs. Nonpoint Source Issues:

EPA. 2008a. Accessed September 21, 2008. http://www.epa.gov/nps/.

EPA. 2008b. Handbook for Developing Watershed Plans to Restore and Protect Our Waters, EPA 841-B-08-002, March 2008. Accessed September 21, 2008: http://www.epa.gov/owow/nps/watershed_handbook/

EPA. 2008c. http://www.epa.gov/owow/funding.html .

USGS. 2006. National Water-Quality Assessment Program, Scientific Investigations Report 2006–5012, County-Level Estimates of Nutrient Inputs to the Land Surface of the Conterminous United States, 1982–2001, Barbara C. Ruddy, David L. Lorenz, and David K. Mueller, U.S. Department of the Interior U.S. Geological Survey, 2006.

Significant Costs of Nutrient Removal:

EPA. 2007. Biological Nutrient Removal Processes and Costs. EPA-823-R-07-002. June 2007. http://www.epa.gov/waterscience/criteria/nutrient/files/bio-removal.pdf

EPA. 2008. Municipal Nutrient Removal Technologies Reference Document, Volume 1 – Technical Report. EPA 832-R-08-006. Washington, D.C.

Sustainability:

Cordell, Dan, Dr Tina Schmid-Neset, Dr Stuart White and Dr Jan-Olof Drangert. 2009. Preferred future phosphorus scenarios: A framework for meeting long-term phosphorus needs for global food demand. Proceedings of International Conference on Nutrient Recovery from Wastewater Streams, Vancouver, British Columbia, May 10-13, 2009.

EPA. 2008. 2009-2014 EPA Strategic Plan Change Document. September 30, 2008. U.S. Environmental Protection Agency, Office of the Chief Financial Officer, Washington, D.C.

NACWA. 2009. NACWA Nutrients Issue Paper, Technical Discussion. HDR Engineering, Inc. National Association of Clean Water Agencies, August 2009.

CHAPTER 4.0

TREATMENT TECHNOLOGY

In order to comply with low nitrogen and phosphorus effluent limits, additional nutrient removal is required beyond secondary treatment. These processes are commonly labeled as Advanced Treatment, Enhanced Biological Nutrient Removal, and Tertiary Treatment. These terms are used to describe the additional unit processes and biological treatment modifications required to accomplish the required levels of nutrient removal. Primary and secondary treatment processes only remove a limited fraction of nutrients from wastewater – a portion of the insoluble nitrogen and phosphorus taken out with primary solids and nutrient uptake required for biological growth. Nutrient removal also requires additional energy, chemicals, maintenance materials, and labor which increase the complexity of plant operations and costs.

This chapter summarizes advanced treatment technologies for nutrient removal, including a discussion of nitrogen and phosphorus speciation, and the limits of treatment capabilities. Effluent discharge permitting issues are discussed in Chapter 5.0 and Chapter 6.0 presents example discharge permits with nutrient limits as references.

4.1 Nutrient Species in Wastewater Treatment

Secondary treatment processes do not remove substantial amounts of nutrients from wastewater. Only a fraction of the influent nitrogen and phosphorus is assimilated into the cell mass during biological growth (typically called synthesis) and removed with the excess biomass wasted from the process.

However, secondary treatment processes can be modified to remove additional nutrients from the wastewater. These modifications typically require additional reactor volume, separated environments (zones in the process), and addition of polishing/tertiary processes. In many instances chemical addition is required for biological growth (denitrification, biological phosphorus removal, etc) and/or chemical precipitation of phosphorus.

4.1.1 Nitrogen Species

Nitrogen is present in wastewater as an inorganic or organic form. Inorganic nitrogen species include ammonia, nitrate, nitrite, and to lesser extent other forms (such as nitrous oxide). Organic nitrogen species are many complex molecules, proteins, cell material, and other organic molecules that can be particulate or dissolved.

Nitrogen can be converted from one species to another through biological and chemical action, but is ultimately removed from wastewater in two forms:

- Nitrogen gas ($N_{2(g)}$). Nitrogen in the influent is converted to gaseous nitrogen by converting organic nitrogen to ammonia, ammonia to nitrate (and nitrite) through nitrification, and nitrate (and nitrite) to nitrogen gas through denitrification. Direct conversion of ammonia to nitrogen gas is also possible (see Table 4-1).

- Particulate nitrogen. Natural organic matter contains nitrogen required as an essential element for growth. Particulate nitrogen is then removed with any organic solids including bacteria that are separated from the wastewater in the treatment system.

Table 4-1. Biological Nitrogen Removal and Conversion Processes.

Initial species	Ultimate species	Process
Organic–N	Ammonia, NH_4^+-N	Ammonification - biological conversion of organic nitrogen to ammonia
	Dissolved organic nitrogen (DON)	Decay products from biological treatment and other recalcitrant species that may be of synthetic origin. A fraction of the DON is not biodegradable in the process and appears in the effluent as RDON.
Ammonia, NH_4^+–N	Nitrite, NO_2^--N	Biological ammonia oxidation, first step in nitrification using AOB.
	Nitrate, NO_3^--N	Biological nitrification—in reality the sum of ammonia and nitrite oxidation
	Organic nitrogen (biomass)	Biological uptake during bacterial growth
	Nitrogen gas, $N_{2(g)}$	Anammox—direct oxidation of ammonia to nitrogen gas using nitrate
Nitrate, NO_2^--N	Nitrate, NO_3^--N	Nitrite oxidation using NOB.
	Nitrogen gas, $N_{2(g)}$	Denitrification of nitrite.
Nitrate, NO_3^--N	Nitrogen gas, $N_2(g)$	Biological denitrification
	Organic nitrogen (biomass)	Biological uptake during bacterial growth

Ref. WEF Nutrient Manual of Practice (2009)

Inorganic nitrogen species in Table 4-1 can be measured using standard techniques. Organic nitrogen comprises a large group of compounds and is normally measured as Kjeldahl Nitrogen (Org-N = TKN – NH_3-N). However, there is currently no standard method to determine the biodegradability of the organic species of nitrogen.

The effluent nitrogen is mainly associated with residual dissolved organic compounds and resists conversion/removal through biological processes employed in the wastewater treatment plant. The following definitions are used for nitrogen species in the effluent (Stensel et al., 2009) – see also Figure 4-1:

- Effluent Organic Nitrogen (EON). The effluent organic nitrogen consists of two classes: effluent particulate organic nitrogen (EPON) and effluent dissolved organic nitrogen (EDON). The particulate fraction can be removed with efficient filtration, but the dissolved fraction is difficult to remove.

- Effluent dissolved organic nitrogen (EDON) can be further divided into two fractions; a bioavailable fraction and a recalcitrant fraction:

    - Bioavailable EDON (bEDON) is effluent dissolved organic nitrogen that can be assimilated by bacteria and algae in surface waters. This fraction is of importance in the environment since it may support algal growth.

    - Recalcitrant EDON (rEDON) is effluent dissolved organic nitrogen that is resistant to biological transformation and uptake by algae or other aquatic organisms in surface waters. This fraction is of less important to the environment since it is not expected to encourage algal growth.

The ability to characterize and measure the EDON compounds and determine methods to remove the refractory compounds remains a topic of research.

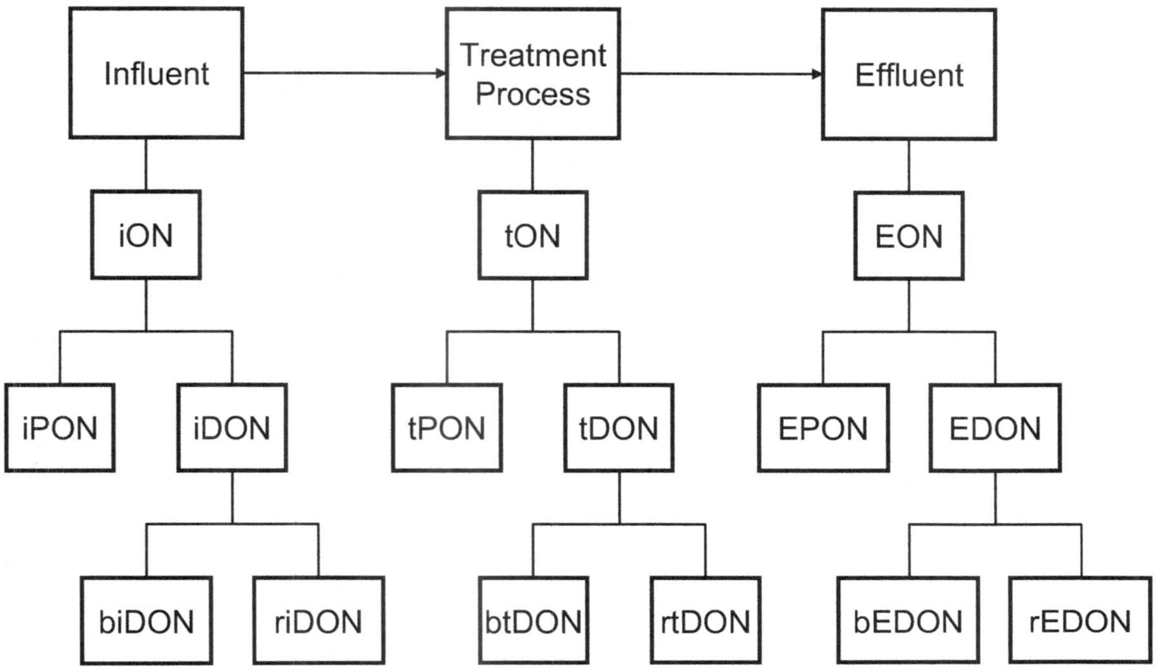

Figure 4-1. Organic Nitrogen Components in Wastewater Influent, Biological Process, and Effluent.
(Stensel et al., 2009).

4.1.2 Phosphorus Species

Phosphorus is present in wastewater as an inorganic or organic form. Inorganic phosphorus species include phosphate, polyphosphates, condensed phosphate, chemical precipitants, etc. Organic phosphorus species are many complex molecules, proteins, cell material, and other organic molecules that can be particulate or dissolved.

Phosphorus can be converted from one species to another through biological and chemical action, but is ultimately removed from wastewater as a particulate in biosolids or other residuals from advanced treatment steps (e.g. tertiary filtration). Table 4-2 summarizes some of the common conversions and reactions that take place during wastewater treatment.

The analytical methods as prescribed in *Standard Methods* can be used to distinguish the various phosphorus fractions in the wastewater. Table 4-3 summarizes the various fractions of phosphorus that are typically measured by these methods. Dissolved phosphorus refers to non-filterable phosphorus, typically measured as the fraction that will pass a 0.45 um filter. These methods do not completely identify the phosphorus species, but measured groups of compounds with respect to their reactivity.

Table 4-2. Phosphorus Species and Reactions.

Species	Common conversion or removal process
Organic-P	Organic phosphorus can be converted to orthophosphate and polyphosphate, some organic species degrade very slowly
Orthophosphate	Most abundant phosphorus species Reactive species in chemical reactions and consumed in biological growth
Polyphosphates	Condensed orthophosphates Possibly reacts with metal salts Can be used for biological growth
Chemical phosphorus	Precipitated phosphates formed by reacting orthophosphate with metal salts, or precipitates as phosphate hydroxides.
Biological phosphorus	Phosphorus incorporated into the biomass for growth Excess phosphorus may accumulate under certain conditions

Table 4-3. Typical Analytical Methods for Phosphorus Measurement.

Fraction	Filter?	Acid?	Heat?	I	II	III	IV	V
A. Total Reactive Phosphorus	No	No	No	Y	N?	N	?	?
B. Total Acid Hydrolyzable Phosphorus[a]	No	Yes	No	N[c]	Y	N?	Y	Y?
C. Total Phosphorus	No	Yes	Yes	Y	Y	Y	Y	Y
D. Total Organic Phosphorus	Calculate as D = C-B-A							
E. Total Non-Reactive Phosphorus	Calculate as E = C-A							
F. Dissolved Reactive Phosphorus	Yes	No	No	Y	N?	N	?	?
G. Dissolved Acid Hydrolyzable Phosphorus[b]	Yes	Yes	No	N[c]	Y	N?	?	?
H. Dissolved Total Phosphorus	Yes	Yes	Yes	Y	Y	N	?	?
J. Dissolved Organic Phosphorus	Calculate as J = H-F-G							
K. Dissolved Non-Reactive Phosphorus	Calculate as K = H-F							
L. Particulate Phosphorus	Calculate as difference of Total and Soluble							

Note (Adapted from Neethling et al. 2007):
I - Phosphate
II - Polyphosphate / Condensed phosphate
III - Organic phosphorus
IV - Chemical phosphorus
V - Adsorbed phosphorus
a. Calculated from measurement that includes (A)+(B).
b. Calculated from measurement that includes (E)+(F).
c. Subtracted from measurement

Phosphorus must be converted to a solid form in order to be removed from wastewater. Two particulate forms are commonly formed in treatment processes (Table 4-2):

- Biomass-associated phosphorus. Biomass typically contains 1.5 to 2.5% phosphorus on a volatile solids dry weight basis. Enhanced biological phosphorus removal (EBPR) processes are designed to favor the growth of phosphorus accumulating organisms (PAOs) that contain significantly more phosphorus than typical biomass. These processes require both aerobic and anaerobic zones and the availability of readily biodegradable organics to sustain the PAO population.

- Chemical phosphorus precipitants. Phosphorus can be converted to a chemical form by adding a metal salt or lime. Phosphorus reacts with chemical precipitating agents to form metal hydroxides or other insoluble species. The precipitated phosphorus particles can then be removed with an efficient solids separation process (clarifiers, filters, membranes, etc.). Metal oxide coated sand and other adsorbents can also be used.

Enhanced wastewater treatment processes employing state-of-the-art phosphorus removal must include effluent polishing filters or other efficient solids separation devices to remove the particulate phosphorus species. However, even with all particulate phosphorus removed, the effluent will still contain residual phosphorus. The residual phosphorus resists conversion/removal through biological processes employed in the wastewater treatment plant. This soluble non-reactive phosphorus is measured analytically as the difference between the dissolved reactive phosphorus (orthophosphate) and the soluble total phosphorus. The non-reactive phosphorus is believed to be primarily organic in nature, but could contain other complexed phosphorus compounds. Similar to EDON, the following definitions are used for phosphorus species in the effluent (adapted from Neethling et al., 2007) – see also Figure 4-2:

- Effluent non-reactive Phosphorus (ENRP). The effluent organic phosphorus consists of two classes: effluent particulate non-reactive phosphorus (EPNRP) and effluent dissolved non-reactive phosphorus (EDNRP). The particulate fraction can be removed with efficient filtration, but the dissolved fraction is difficult to remove.
- Effluent dissolved non-reactive phosphorus (EDNRP) can be further divided into two fractions: a bioavailable fraction and a recalcitrant fraction:
    - Bioavailable EDNRP (bEDNRP) is effluent dissolved non-reactive phosphorus that can be assimilated in surface waters through bacteria and algae uptake. This fraction is of importance in the environment since it will support algal growth.
    - Recalcitrant EDNRP (rEDNRP) is effluent dissolved non-reactive phosphorus that is resistant to biological transformation and uptake by algae and bacteria in surface waters. This fraction is of less importance to the environment since it is not expected to encourage algal growth.

There is currently no standard procedure to measure bioavailable or recalcitrant phosphorus.

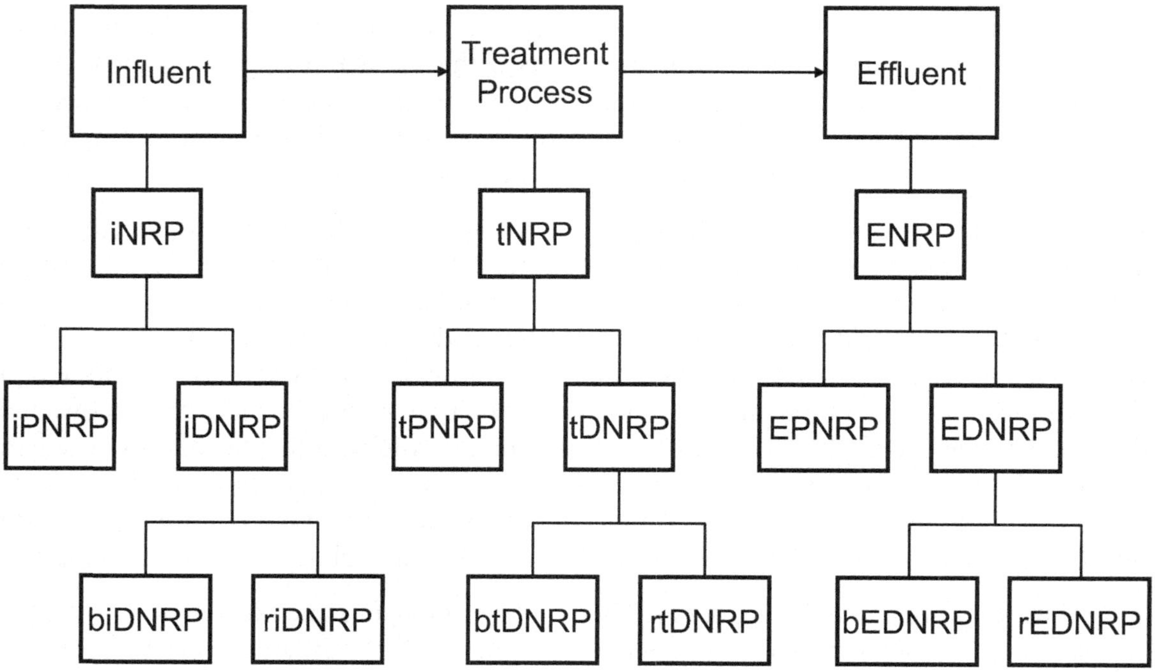

Figure 4-2. Non-Reactive Phosphorus Components in Wastewater Influent, Biological Process, and Effluent.
Adapted from Neethling et al, 2007, and then Stensel et al., 2009.

4.2 Available Treatment Technologies and Capabilities

Treatment technologies for nutrient removal can be grouped as follows:

- Target nutrient removed: nitrogen and/or phosphorus. Nitrogen removal processes are mainly biological in nature, while phosphorus removal processes can be either biological or physical-chemical.
- Mechanism of removal: physical, chemical, and biological.
- Performance and selection of nutrient removal processes to achieve various treatment levels.

4.2.1 Nitrogen Removal Technologies

Figure 4-3 lists the nitrogen removal technologies used in wastewater treatment. Biological processes are the most commonly employed methods. The conventional suspended growth (activated sludge) process can be adapted to achieve nitrification and denitrification in different zones in the process. Several fixed film processes are available – these can be sequential or integrated with BOD removal. Physical/chemical processes are rarely used in municipal treatment systems and are more common in industrial applications.

The various nitrogen removal treatment schemes are illustrated in Figure 4-4.

- Separate processes shown in Schematics 1 to 7 can be designed to remove a particular nitrogen species. Some modification to these basic processes can achieve multiple objectives (for example, some nitrification and denitrification in a biological aerated filter (BAF)). These processes can also be added to existing facilities to achieve nitrogen removal. Fixed film processes typically fall into this category.

- Staged processes shown in Schematics 8 to 12 achieve sequential nitrification followed by denitrification. These processes use separate basins to create aerobic and anoxic conditions for nitrification and denitrification. Denitrification is through substrate level or endogenous level. These are primarily suspended growth processes. Many other process arrangements are possible with a suspended growth system to achieve sequential nitrification and denitrification.
- Integrated nitrification/denitrification processes shown in Schematics 13 and 14 illustrate processes where nitrification and denitrification occur within the same reactor by manipulating the environmental conditions. These processes are typically low rate and have demonstrated very good removal efficiencies.

Biological nitrogen removal technologies do not remove all the organic nitrogen and even produce some dissolved organic nitrogen byproducts. Consequently, the effluent from these processes still contains dissolved organic nitrogen (EDON fraction) that is not susceptible to degradation in biological wastewater treatment.

Membrane processes such as nanofiltration (NF) and reverse osmosis (RO) can be used to remove dissolved nitrogen and phosphorus compounds to an extent beyond that of biological processes. NF and RO are specifically effective in removing very small particulate forms of nitrogen, large soluble molecules, and ionic species, respectively. However, these processes are not commercially practiced strictly for enhanced nutrient removal. They are sometimes incorporated in water reclamation systems which can remove additional nutrients as a side-benefit.

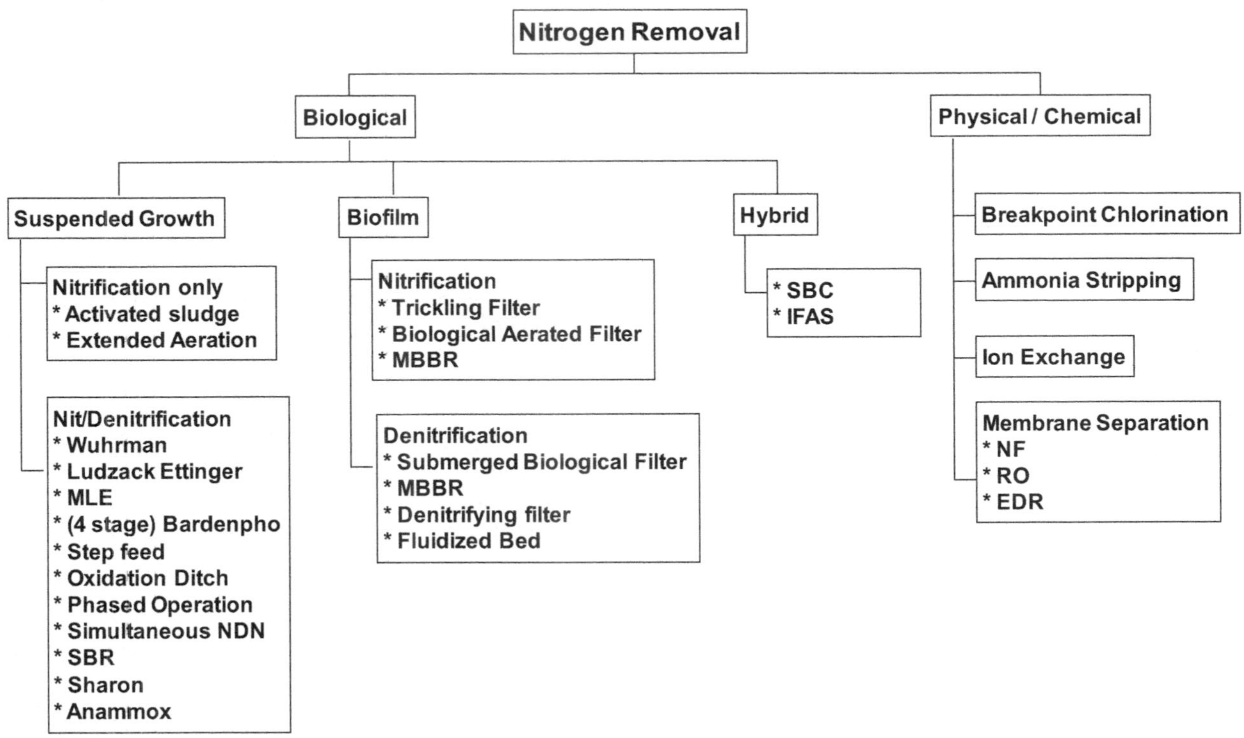

Figure 4-3. Processes Used for Nitrogen Removal.
Reprinted with permission from *Nutrient Removal,* Manual of Practice No. 34, Copyright © 2010, Water Environment Federation, Alexandria, VA. www.wef.org.

No.	Process Schematic	Process	Susbrate Level Denitrification	Endogenous Denitrification	Integrated Nit / Denit	Separate Nitrification	Separate Denitrification
1	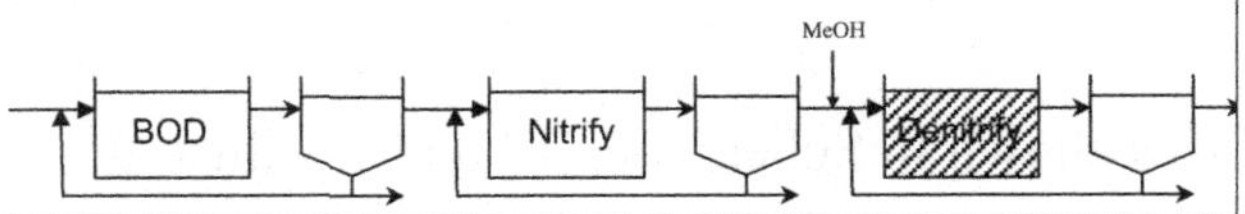	Activated Sludge[a]				X	X
2	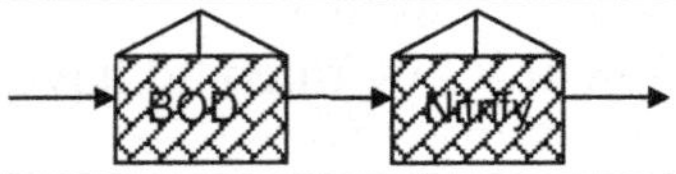	Trickling Filter[a]				X	
3	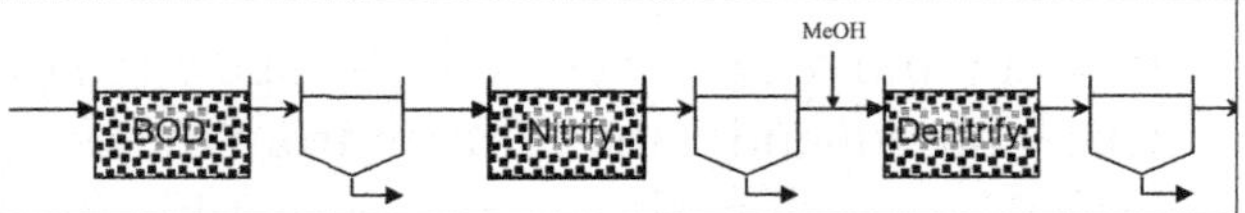	MBBR[a]				X	X
4	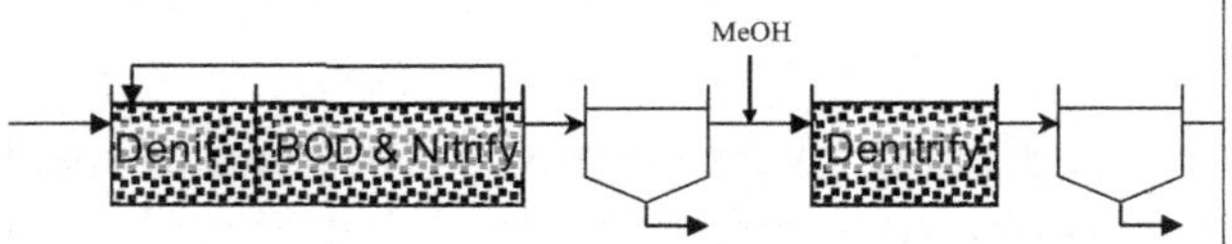	MBBR[a]	X	X		X	X
5	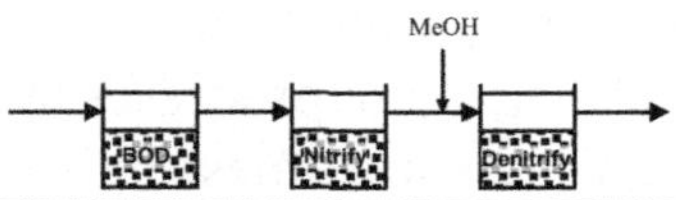	BAF[a]				X	X
6	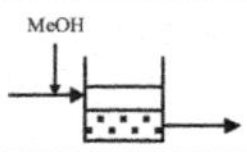	Effluent filter[a]					X
7	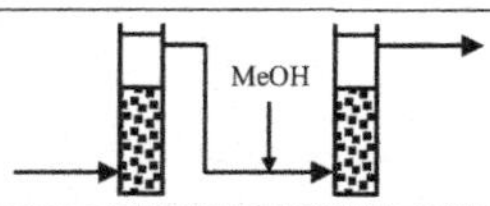	Fluidized bed				X	X
8	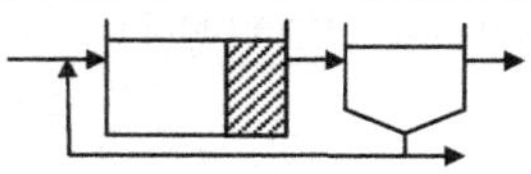	Wuhrman		X			
9	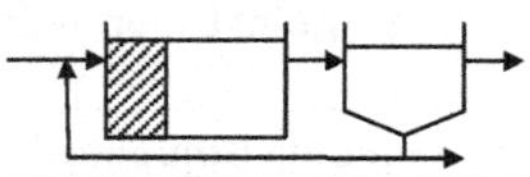	Ludzack Ettinger	X				
10	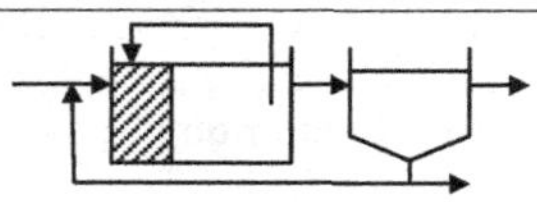	Modified Ludzack Ettinger (MLE)	X				
11	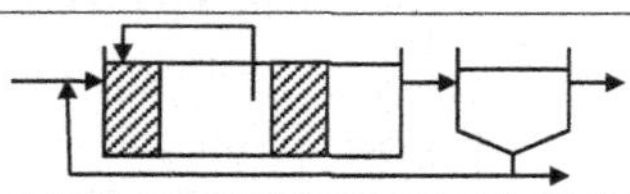	Bardenpho (4-stage)	X	X			
12	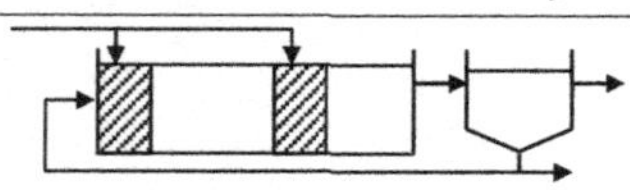	Step feed	X	X			
13	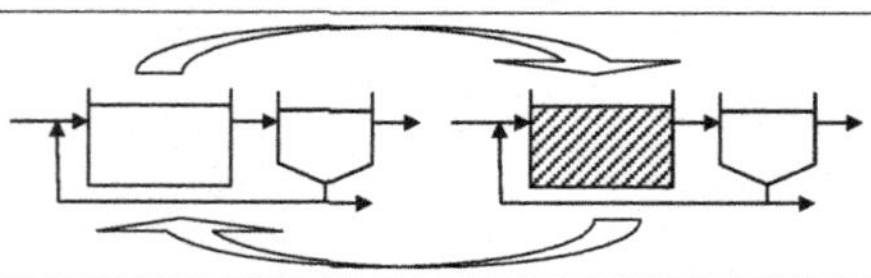	Phased operation[b]			X		
14	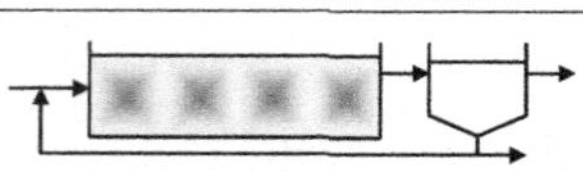	Simultaneous NDN			X		

No.	Process Schematic	Process	Susbrate Level Denitrificat ion	Endogeno us Denitrificat	Integrated Nit / Denit	Separate Nitrificatio n	Separate Denitrificat ion
15	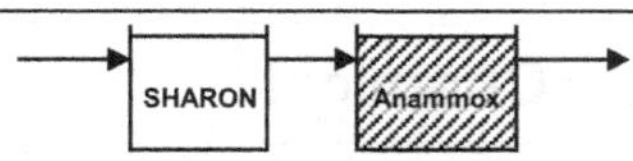	Anammox			X		

Figure 4-4. Typical Biological Nitrogen Removal Schemes.

Notes:
Separate processes indicate those that can be added to another unit process
a. As a stand alone process without zones or other process schemes. Can be added for single objective (ex. Nitrification)
b. Phased operation to generate sequential nitrification and denitrification in the same reactor (example Biodenitro, SBR)
c. Simultaneous operation to achieve nitrification and denitrification at the same time
d. Hybrid processes are considered a variation on the activated sludge process and can be incorporated into activated sludge

4.2.2 Phosphorus Removal Technologies

Phosphorus removal processes can be grouped into physical/chemical and biological processes (Figure 4-5). Both physical/chemical and biological phosphorus processes are used extensively and can be adapted to most secondary treatment plants by retrofitting existing facilities or adding new facilities. Biological processes are often designed for both nitrogen and phosphorus removal. However, both biological nitrogen and biological phosphorus removal requires readily biodegradable organics to achieve efficient denitrification and phosphorus uptake. An external source of readily biodegradable organics is often required. This can be provided by chemical addition (methanol, acetic acid, etc.) or by providing a sludge fermenter to produce organics on site from the primary treatment solids.

All phosphorus removal occurs by converting the soluble phosphorus to a particulate form (biomass or chemical) and then separating the particles from the liquid. Particle separation is achieved in clarifiers, filters, or membrane processes. In order to achieve very low phosphorus concentrations, essentially all particles must be removed, therefore requiring very efficient solids separation.

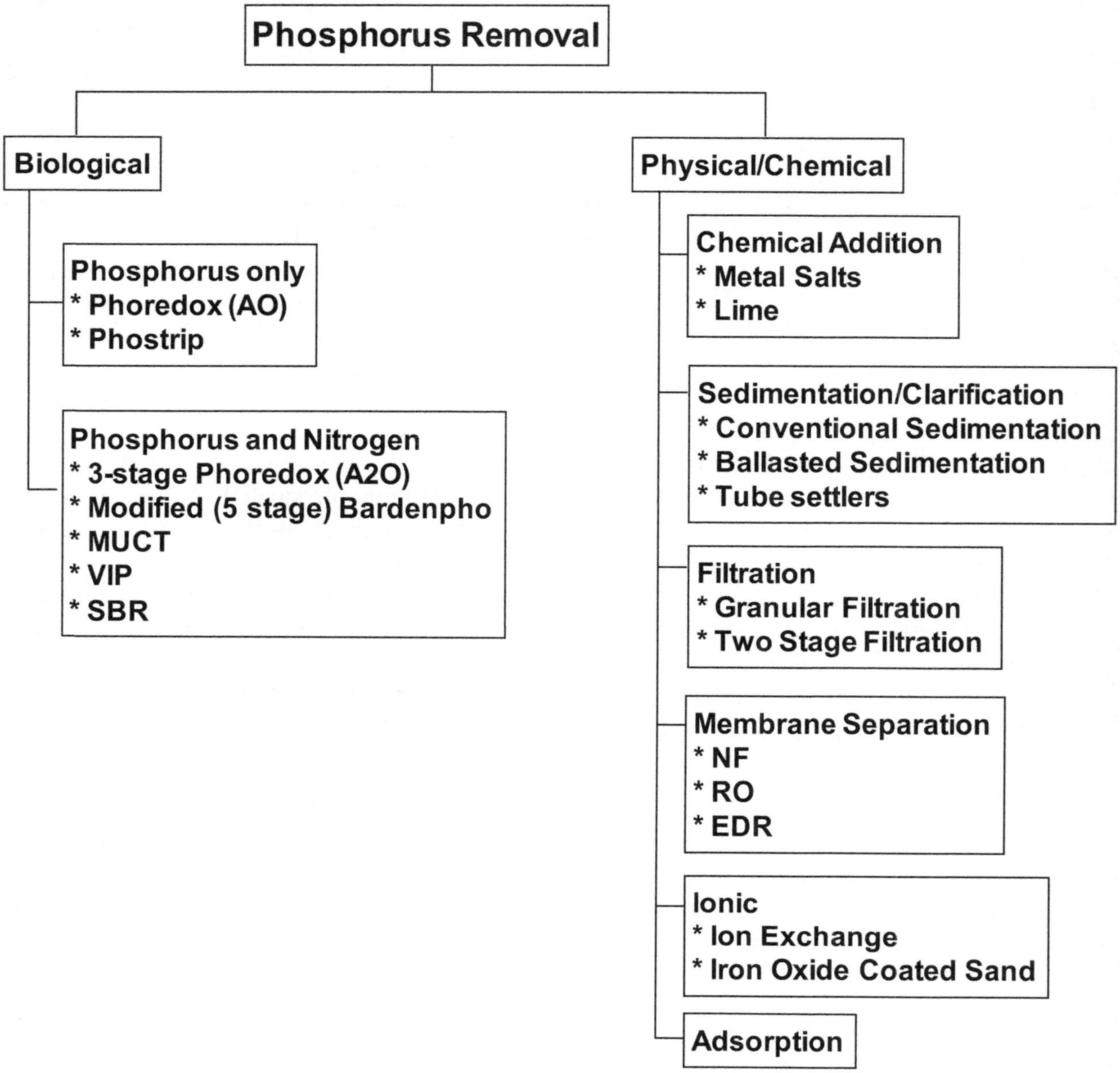

Figure 4-5. Processes Used for Phosphorus Removal.
Reprinted with permission from *Nutrient Removal,* Manual of Practice No. 34, Copyright © 2010, Water Environment Federation, Alexandria, VA. www.wef.org.

Process Schematic	Process	Biological P removal	Chemical P Removal	P Removal only	N&P Removal	Tertiary
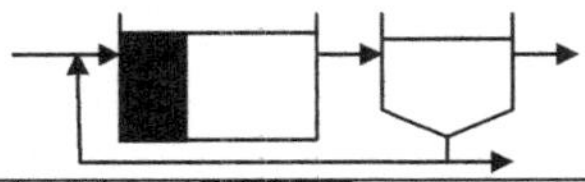	Phoredox (AO)	X		X		
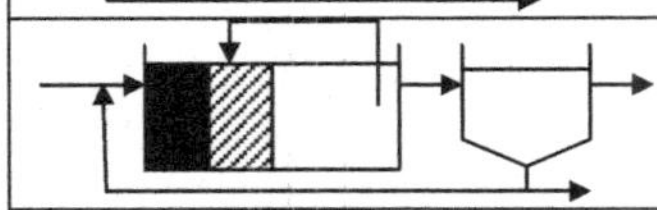	3-stage Phoredox (A2O)	X			X	
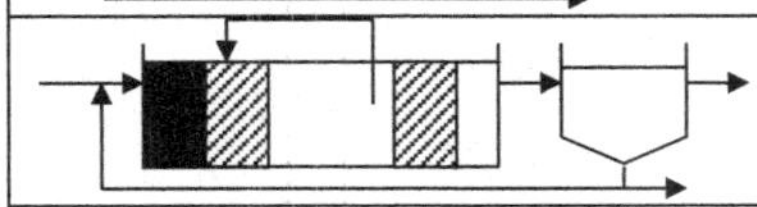	Modified (5 stage) Bardenpho	X			X	
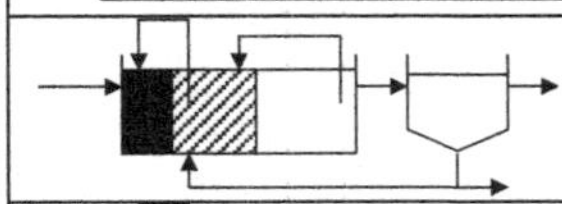	UCT	X			X	
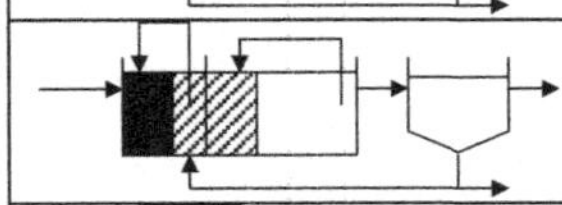	mUCT	X			X	
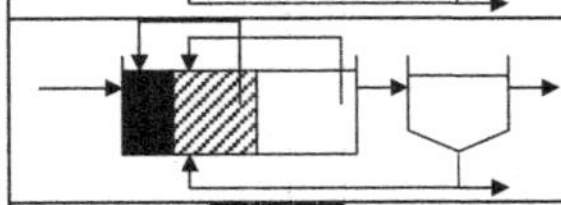	VIP	X			X	
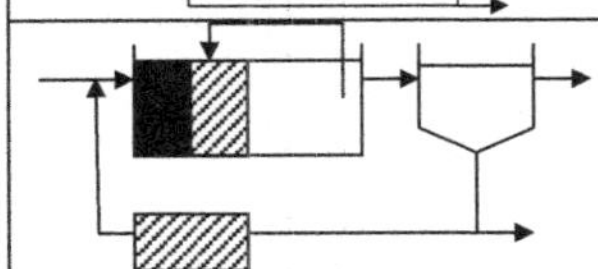	Johannesburg Process	X			X	
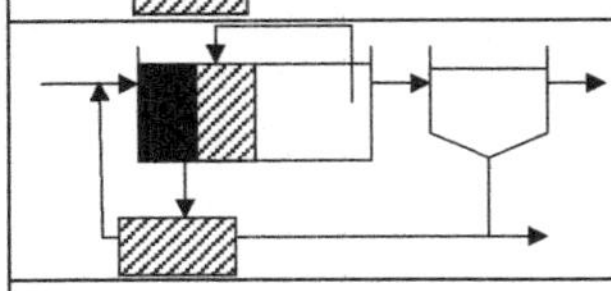	Modified Johannesburg Process	X			X	
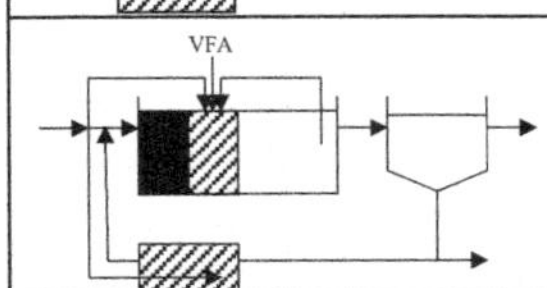	West Bank Process	X			X	
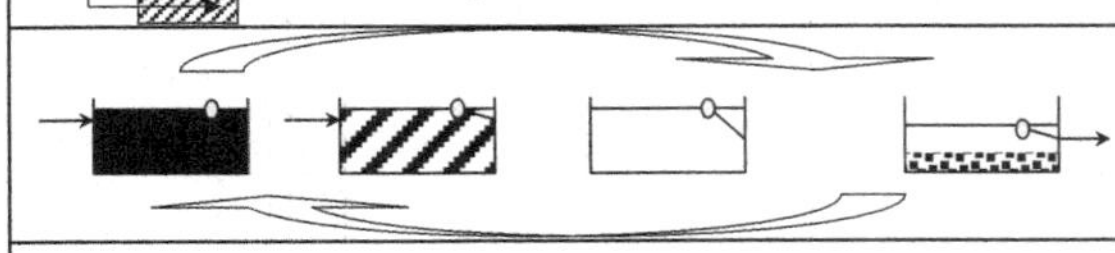	Sequencing Batch Reactor	X			X	
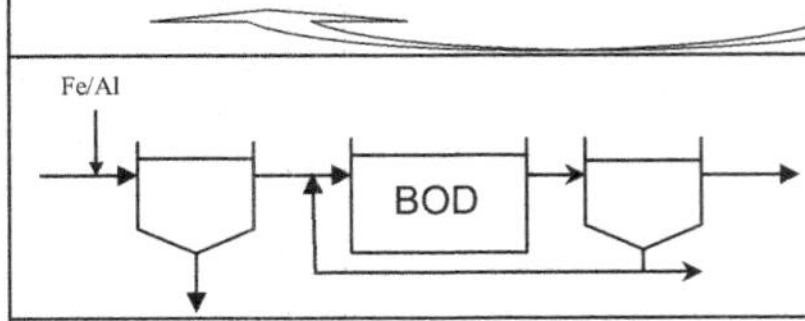	Primary Chemical Secondary		X	X		
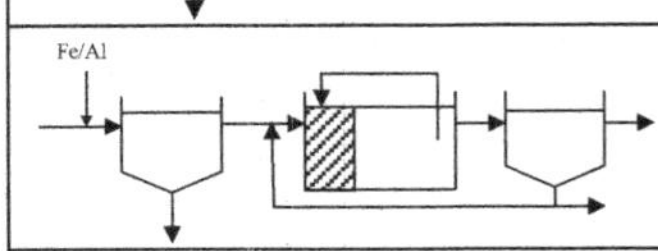	Primary Chemical Any NDN process		X		X	

Process Schematic	Process	Biological P removal	Chemical P Removal	P Removal only	N&P Removal	Tertiary
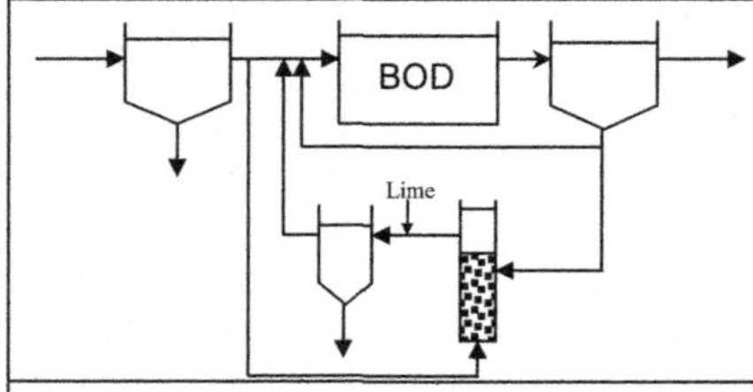	Phostrip	X	X	X		
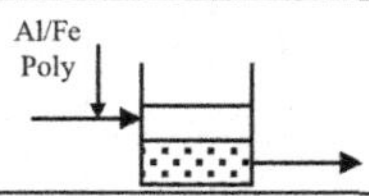	Direct Filtration		X	X		X
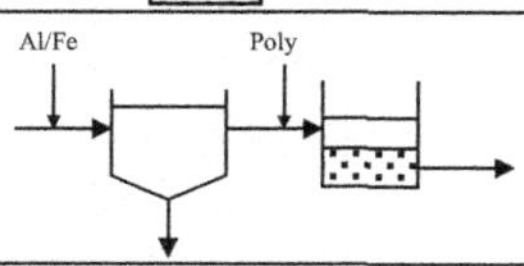	Sedimentation (typ) Filtration		X	X		X
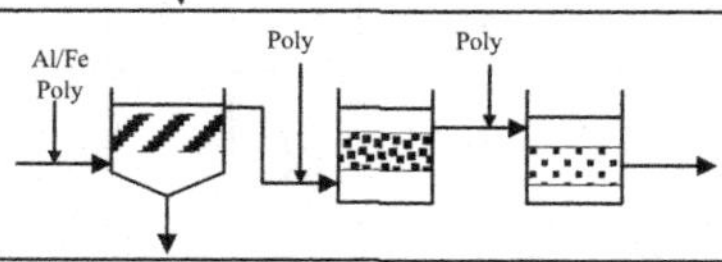	Enhanced Sedimentation/Filtrati on (Trident High Solids)		X	X		X
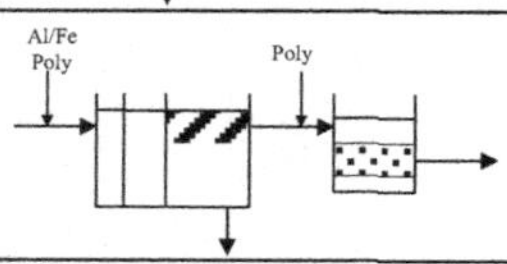	Ballasted Sedimentation/ Filtration (Actiflo/CoMag)		X	X		X
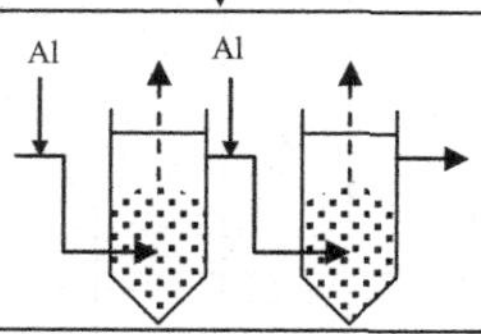	Two Stage Filtration (Dual filtration)		X	X		X
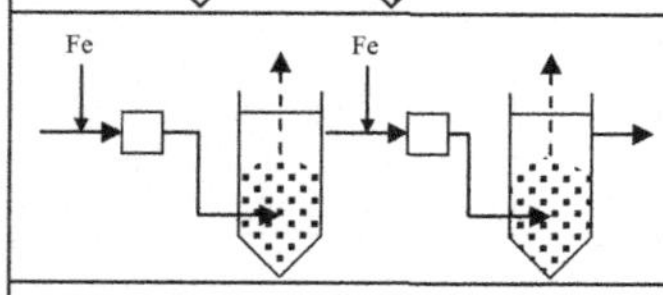	Iron Oxide Coated Filtration (BluePro)		X	X		X
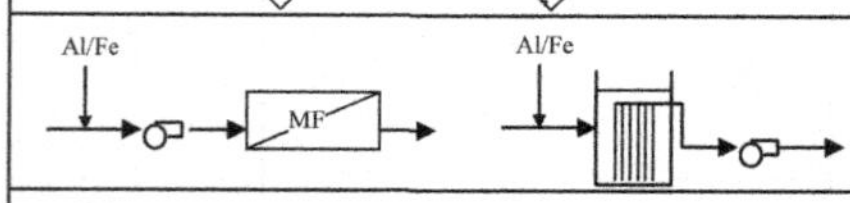	Microfiltration		X	X		X
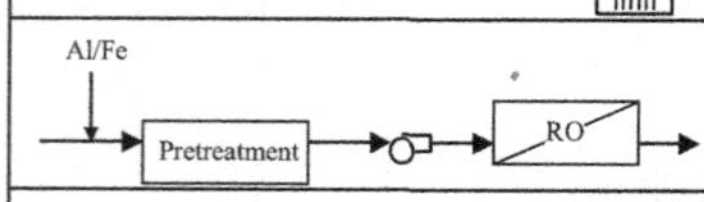	Reverse Osmosis		X	X		X
Anaerobic Anoxic Aerobic						

Figure 4-6. Typical Phosphorus Removal Schemes.

4.2.3 Technology Capabilities

Nutrient removal processes can be classified in three "levels" of effluent concentration. (WEF, 2009). These levels are used for convenience to describe the technologies and their capabilities and do not necessarily apply to industrial waste treatment systems. Industrial wastewater can have higher nutrient concentrations and forms that are more difficult to treat. These levels may not then be attainable for some industrial WWTPs and municipal WWTPs that have a substantial industrial waste contribution.

Some regions (for example Chesapeake Bay) or states (for example Maryland) have proposed levels of treatment to be accomplished, using common terminology such as BNR (biological nutrient removal) or ENR (enhanced nutrient removal) to describe a nutrient level. Instead of selecting a particular set of nutrient levels that has specific significance to a particular region or state, a general set of treatment levels or objectives can be used to describe process selection. These levels, shown in Table 4-4, form the basis for discussion and identifying process performance for treatment technologies in this section.

Table 4-4. Municipal Wastewater Nutrient Removal Treatment Levels[1,2]

Level	Total Nitrogen	Total Phosphorus	Comment
1	8	1	Achievable with conventional nutrient removal technologies. Chemical addition or filtration is typically not required.
2	3	0.1	Enhanced removal requires tertiary treatment and chemical addition to achieve low cocentrations.
3	1	0.01	Requires state-of-the-art technology and enhanced/optimized, treatment operation. May or may not be feasible, especially to simultaneously achieve both the very low N and P levels.

[1] Use modifier "N" or "P" to denote limits for nitrogen or phosphorus only. For example, Level 2P is only phosphorus level of 0.1 mg/L TP, with no nitrogen level.

[2] The expected nutrient treatment removal levels and associated effluent concentrations vary widely according to the averaging period and/or performance statistic employed.

This section discusses the features of technologies or processes that can achieve these limits. It also includes a section on the refractory nutrients that limits the ability of all the technologies.

4.2.3.1 Level 1 – Conventional Municipal Nutrient Removal

Level 1 represents the capability of conventional municipal nutrient removal processes. A conventional process is typically a modification of a secondary process or series of processes. These processes typically do not include chemical addition to supplement readily biodegradable BOD but may include single stage chemical phosphorus removal. Filtration is not typically required to for Level 1 – conventional nutrient removal.

While the performance level depends on the process design, the influent composition of the wastewater, and in particular, the availability of readily biodegradable organics, most municipal treatment systems can control effluent nitrogen to Level 1.

Level 1 phosphorus removal is achievable with single stage phosphorus removal processes (chemically or biologically) under typical conditions. Neethling et al. (2005) reported that conventional biological phosphorus removal can achieve 0.5 mg/L TP under favorable conditions.

Although normally not required, effluent filtration adds robustness to the treatment plant performance by further reducing particulate nutrients (both nitrogen and especially phosphorus). Since phosphorus is concentrated in the suspended solids (TSS phosphorus content in phosphorus removal plants increase from the typical 1.5 to 2.5% to the 3 to 6% and in some cases even higher). At 5 % phosphorus content, a 20 mg/L effluent TSS will contain 1 mg/L particulate phosphorus and make it difficult to achieve Level 1.

4.2.3.2 Level 2 – Enhanced Nutrient Removal (ENR)

Level 2 represents the capability of enhanced nutrient removal processes. An enhanced nutrient removal process is an upgrade of the conventional nutrient removal technology to include additional reliability and performance enhancements. ENR processes often include multiple upgrades to remove each nutrient of interest. These upgrades typically include chemical addition to supplement readily biodegradable BOD for nitrogen removal (denitrification) and possibly for biological phosphorus removal. Filtration is required to achieve Level 2.

ENR processes meet lower nitrogen effluent concentrations by reducing ammonia and nitrate to very low levels. The inorganic nitrogen must be reduced to 1 to 2 mg/L to meet Level 2 since a portion of DON remains in the effluent (typically 1-1.5 mg/L). Ammonia is oxidized to a very low concentration (below 0.1 mg/L in many cases) which means that oxidized nitrogen (nitrate plus nitrite) must be reduced to 1 to 2 mg/L.

Supplemental readily biodegradable carbon is required to reduce oxides of nitrogen to the 1 to 2 mg/L required to meet Level 2 levels. In addition, a tertiary denitrification stage is required. Some 4-stage NDN processes can meet this limit by adding an external carbon source to the second anoxic stage. Tertiary denitrification processes can be used to reduce the nitrate as needed.

Level 2 phosphorus removal requires a two stage phosphorus removal process and specifically a tertiary treatment step. The first stage can be a chemically or biologically conventional phosphorus removal process. The tertiary stage must include chemical addition and a very effective solids removal stage.

4.2.3.3 Level 3 – Best Achievable Performance

Level 3 represents the maximum potential capability that state-of-the-art technology can achieve or the best achievable performance. It is debatable what the best performance is, since additional barriers or treatment steps can always be added. However, these tertiary and beyond treatment processes are rarely cost-effective. Also, performance depends on the local conditions (weather), process implementation, characteristics of the wastewater, treatment target, external factors (construction, extreme weather), and operator skill to a name a few.

Process configurations that the lowest technology achievable levels typically contain certain features:

- Provide multiple barriers for each nutrient to be removed
- Eliminates recycle stream impacts, for example from solids treatment
- Provide extremely efficient solids separation (typically microfiltration, two stage media filtration, clarifier/filtration, etc).

- Minimize fluctuations in influent flow and composition by flow equalization and/or dampening peak flows in the collection system. Avoid industrial discharges that can impact performance.
- Provide chemical feed (supplemental carbon source) and metal salts for phosphorus removal.
- Instrumentation and control to provide automation of critical process functions such as aeration control, chemical feed, flow pacing, etc. Online instrumentation to monitor process performance.

4.2.3.4 Recalcitrant Dissolved Organic Nitrogen (RDON)

The Recalcitrant Dissolved Organic Nitrogen (also called Refractory Dissolved Organic Nitrogen) or RDON for short is the fraction of nitrogen that is resistant to biological treatment. Figure 4-1 shows the different dissolved organic nitrogen (DON) fractions in wastewater treatment. The figure identifies influent (iDON), treatment process (tDON), and effluent DON (EDON) associated fractions.

Changes in the nature of DON during the biological treatment process are still poorly understood. DON composition can also be influenced by certain nitrogen-containing industrial wastewater. Some DON is degraded by the biomass in the process while the endogenous decay during treatment produces additional DON. The DON produced during treatment is a function of the process: type of process, sludge age, operational settings, solids processing, etc. These biological processes cause a change in the nature of the DON at various locations in the treatment plant.

The degradation of the DON during biological treatment is dependent on the environmental and ecological conditions as well as its specific composition. Two definitions are important: Biodegradability refers to the ability of the microorganisms in a treatment plant to use/transform the DON in the wastewater under the conditions (pH, DO, temperature, contact time, etc.) in the treatment plant. The biodegradable DON (btDON) (Figure 4-1) is expected to be very low since the biology would remove this fraction during treatment. Bioavailability is a broader term that captures the ability of bacteria, algae and other organisms to use DON to support growth under natural conditions (temperature, salinity, sunlight exposure, biological, long time periods, etc.). The bioavailable DON in the treatment plant effluent is the bEDON (see Figure 4-1).

By definition, the rtDON fraction is not biodegradable. Treatment techniques such as oxidation, chemical coagulation, electric pulse or adsorption have not been commercially demonstrated to reduce the rtDON fraction or convert it into a biodegradable constituent. Khan (2009) and Parkin and McCarthy (1981) suggested that 40-60% of the influent DON is recalcitrant. In addition, Parkin and McCarthy (1981) also suggested that 20% of the EDON is produced by the biological process.

Stensel et al. (2009) summarized reported literature on this subject and concluded that EDON concentrations from BNR plants range from 0.10 to 2.80 mg/L with 50th and 90th percentiles at 1.2 and 2.1 mg/L EDON respectively. Results from Pagilla (2007) found that 68% of 188 facilities in Maryland and Virginia had EDON concentrations below 1.0 mg/L. EDON from industrial facilities has been significantly higher – values at 15 mg/L and above have been measured.

4.2.3.5 Recalcitrant Dissolved Nonreactive Phosphorus (RDNRP)

The Recalcitrant or Refractory Dissolved Nonreactive Phosphorus (RDNRP) is the fraction of phosphorus that is resistant to biological treatment. Figure 4-2 shows the different dissolved Nonreactive Phosphorus (DNRP) fractions in wastewater treatment. The figure identifies influent (iDNRP), treatment process (tDNRP), and effluent DNRP (EDNRP) associated fractions.

The nonreactive phosphorus is presumed to be comprised mainly of organic matter and therefore represents a corollary to the DON fraction discussed above. Research to understand the nature of this non-reactive phosphorus is ongoing.

Changes in DNRP composition during the biological treatment process are still poorly understood. Some DNRP is degraded by the biomass in the treatment process while the endogenous decay during treatment produces additional DNRP. The DNRP produced during treatment is a function of the process: type of process, sludge age, operational settings, solids processing, etc. These biological processes cause a change in the nature of the DNRP at various locations in the treatment plant.

The degradation of the DNRP during biological treatment is dependent on the environmental and ecological conditions. Two definitions are important: Biodegradability refers to the ability of the microorganisms in a treatment plant to use/transform the DNRP in the wastewater under the conditions (pH, DO, temperature, contact time, etc.) in the treatment plant. The biodegradable DNRP in the treatment plant is btDNRP (Figure 4-2) is expected to be very low in a typical wastewater treatment plant since the biology would remove this fraction during treatment. Bioavailability is a broader term that captures the ability of bacteria, algae and other organisms to use DNRP to support growth under natural conditions (temperature, salinity, sunlight exposure, biological, long time periods, etc.). The bioavailable DNRP in the treatment plant effluent is the bEDNRP (see Figure 4-2).

At this point in time, treatment techniques have not been commercially demonstrated to reduce the rtDNRP fraction or convert it into a biodegradable constituent. Techniques such as oxidation, chemical coagulation, electric pulse, adsorption, and others are potential ways to make the rtDNRP biodegradable. The DNRP fraction is expected to follow trends similar to the DON and that 40-60% of the influent DNRP is possibly recalcitrant.

Benisch et al. (2007) reported residual EDNRP concentrations from advanced tertiary pilot studies in Coeur d'Alene, ID, and found a residual of 11 to 15 μg/L remain in the effluent. A similar but shorter pilot study of three tertiary pilot units in Marlborough, MA, for a domestic wastewater with nearly 50% commercial and industrial components revealed intermittent recalcitrant phosphorus concentrations up to 500 μg/L (Lancaster and Madden, 2008). It is suspected that recalcitrant phosphorus is mostly, if not entirely, comprised of dissolved organic phosphorus (Neethling, et al., 2007). Some industrial effluents can contain 1,000-2,000 μg/L DNRP.

4.3 Operational Performance

4.3.1 Technology Performance Statistics as Descriptor of Plant Performance

Neethling et al. (2009) introduced a method for using a statistical approach to describe process performance. In this approach, the treatment plant or technology performance is tied to the statistical rank to express the probability of achieving a certain performance. Building on this statistical approach, the term Technology Performance Statistic (TPS) was used at a Water Environment Federation Technical Exhibition and Conference (WEFTEC) workshop (WEF/WERF, 2009) to assess the performance of full scale treatment plants.

The Technology Performance Statistics (TPS) describes the performance of a technology or process or plant under specific conditions (see Table 4-6). The TPS is determined from performance data and is linked to the operational conditions during which the data were collected (pilot, full scale, summer, winter, excess capacity available, SRT, etc). The conditions must also include external factors that impact the technology, industrial loadings, seasonality, absence of recycle streams, etc. In addition, the TPS established using past performance, is tied to the treatment objectives or permit limits.

Table 4-5. Technology Performance Statistics – Special Conditions to Note Regarding Dataset.

Condition	Report	Significance
Treatment goal	Numerical value and period	The treatment goal is typically the regulatory permit limit. In some cases, the goal is lower than the permit. This represents the main target for the operator. Operators can choose to reduce chemicals, energy consumption, etc. to increase efficiency.
Data source	Data source, period, frequency	Regulatory controlled data (permit reports) are the most commonly used data source. Data is assumed to be from a certified laboratory. The dataset duration (number years) and frequency of data collected (samples per period) should be noted. Averaging of data (monthly reports) can be used under certain circumstances; daily data is commonly used.
Season or period	Season	The data period of data collection impacts the conclusion regarding performance. If the dataset is less than a year, no firm conclusions regarding annual operation can be drawn (unless the plant experiences no seasonal changes).
Exclusions	Conditions or data excluded	In some cases a known problem may skew the data (construction, for example). This should not be used to eliminate "poor" or "good" data.
Treatment capacity	Load and capacity	Plants typically operate below their design capacity.
Scale	Pilot, bench, full, etc.	The scale of the process impacts the ability to control the performance. Plants (pilot or other) that have the ability to fully control the influent composition or flow will typically perform better.
Solids processing	Type and recycle stream management	Recycle streams from solids processing could impact performance of nutrient removal plants attempting to achieve low limits.
Special conditions	Special conditions	Special conditions that applies to the application. Industrial contributions, extreme cold or warm conditions, seasonal visitors or slug loads, etc.

TPS values can be used to describe the performance of a treatment process or technology as summarized in Table.

TPS-14d: This is the Best Achievable Performance and represents the performance of the technology under the optimal or best operating conditions. The optimal conditions could be carefully controlled laboratory conditions with defined, treatable influents. For full scale performance, the lowest TPS represent the lowest concentrations (best performance) observed under the conditions experienced at the treatment plant. The 14-day period is selected as a reasonable period to demonstrate the expected sustained best performance of the process.

The TPS-14d is not an appropriate permitting level since the treatment process can only sustain this limit for a relatively short period of time (up to two weeks per year). The TPS-14d value is exceeded 50 weeks per year or more than 96% of the time.

TPS-50%: This is the Average Process Performance and is represented by the median performance. This is indicative of the annual average performance achievable under the operating conditions. The process performance exceeds the TPS-50% half the time or six months in a year.

TPS-95%: This is the Reliable Process Performance and represents a concentration that can be achieved 95% of the time. On a monthly basis, this value is exceeded three times in a five-year period (three months out of 60 months or 5% of the time). This then should represent a reasonable descriptor of reliable performance. For permit compliance a utility can use the performance statistics to determine the reliability required to meet their treatment goals in terms of the operator proficiency, process performance, and acceptance of risk.

Table 4-6. Application of Key Technology Performance Statistic Values.

Limit	Notation	Determine	Interpretation	Performance Implication
Best Achievable Performance	TPS-14d	3.84th percentile	The best performance possible with the technology under the optimal or best operating conditions. This represents the LOT (Limit of Technology).	This limit will be exceeded 96% of the time.
Average Technology Achievable Limit	TPS-50%	50th percentile	This represents a measure of the concentration that was achieved on a statistical annual average basis.	As the median performance, the process exceeds this 6 times per year.[2]
Reliable Technology Achievable Limit	TPS-95%	95th percentile	This represents the concentration that can be achieved reliably by the technology.	This limit is exceeded 0.6 times[2] per year – 3 times in a 5 year period.

Note:

1. Represent the lowest 14-d running average
2. Times = months as typically reported in permits

4.3.2 Performance of Facilities

A number of studies and reports have been published on the performance of "exemplary" municipal nutrient removal facilities. The facilities were typically selected because their demonstrated past performance indicated that the plant was well operated and maintained and had sufficient data to produce a reasonable record.

4.3.2.1 Sources for Data

The following sources of information were used:

- *EPA Municipal Nutrient Removal Technologies Reference Document (2008)*. This document was peer reviewed and published by EPA.
- EPA Region 10 published a report on *Advanced Wastewater Treatment to Achieve Low Concentrations of Phosphorus* (Ragsdale, 2007)
- Spokane River Dischargers authorized a study: *Evaluation of Exemplary WWTPs Practicing High Removal of Phosphorus* (Dave Reynolds, CH2MHill and Dave Clark, HDR, November 21, 2005).

- Esvelt Environmental Engineering (2006) technical memorandum summarizes the data from several Colorado plants achieving very low phosphorus concentrations. Performance data are presented as individual data and summarized statistically.
- WEF/WERF Workshop (2008 and 2009). Operators from 22 different plants participated in this workshop to share operating data and experiences. Daily data from the plants were statistically analyzed and reported in terms of the 3.84th, 50th and 95th percentiles.

These references based their analysis on plant data collected by operational staff and typically reported as part of the NPDES permit requirements. The reports were generally not peer reviewed, but the data presentation is considered reliable as collected by qualified technicians.

4.3.2.2 Summary of Plant Performance

The data reported in these studies were reanalyzed to determine the Technology Achievable Levels indicated above. It was not possible to determine all the three statistical levels from each report, since the original data were not always reported. In some cases reasonable estimate of the TPS values could be made. For example, Esvelt (2006) does not report daily values, but only weekly average results. Since this period is equal or less than the TPS-14d (2 week) value, the 3.84th percentile can be used to determine the TPS-14d value. Similarly, the maximum month value could be used to estimate the TPS-95% value from the 95th percentile.

Comparison of Plant Performance from Different Reports

Several of the documents above investigated the same plants; although typically using slightly different methods and typically for a different period. Figure 4-10 shows the results from total phosphorus from several plants repeated in the different reports. The plants are sorted approximately from low to high and grouped together.

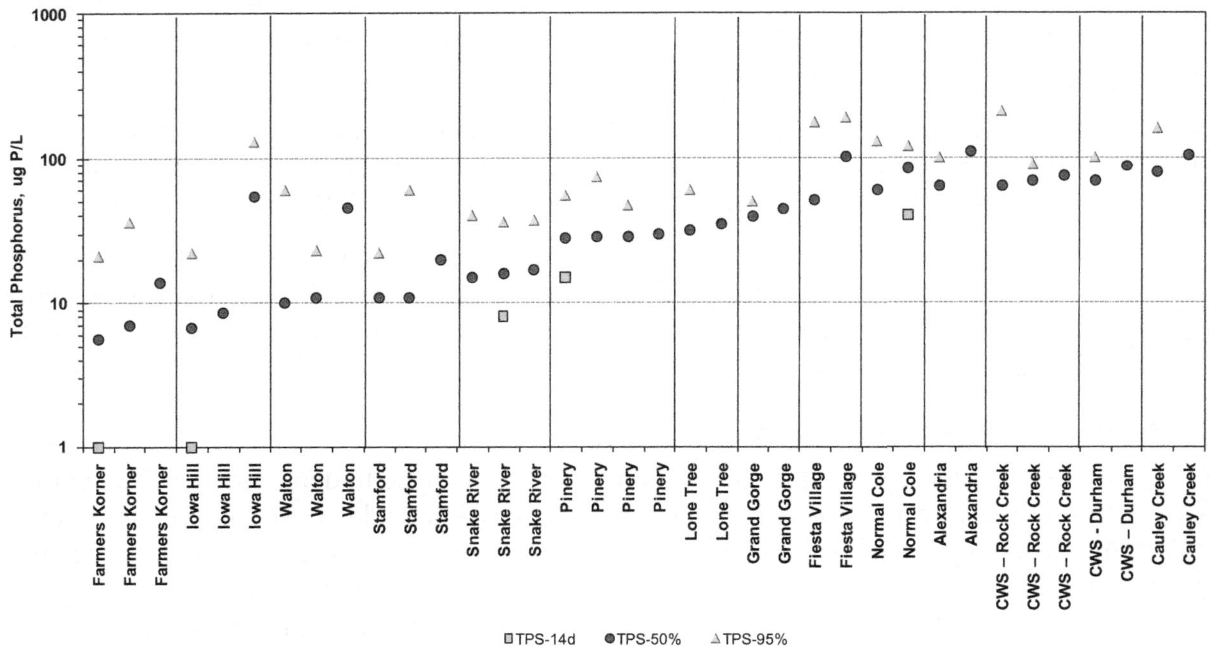

Figure 4-7. Phosphorus Concentrations Reported in Different Studies for the Same Plants (note LOG scale is used).

The results in Figure 4-7 show that:

- In some cases (Farmers Corner, Iowa Hill, Walton, Fiesta Village), the average performance varied from period to period widely. Average performance for Iowa Hill averaged below 10 µg/L for two studies, but 55 µg/L in the third evaluation. Walton averaged about10 µg/L for two studies, but increased to 46 µg/L in the third evaluation.
- The different studies (periods) indicated different results. However, even though the results change for the study period, the general trend remains similar; i.e. the plant performance remains somewhat consistent when compared to other plants. For example, Pinery shows virtually the same performance for four reported studies. However, the three reported TPS-95% values, shows considerable variation. This indicates that even though the average performance remained similar, the maximum month performance for the three periods were quite different.

These findings are important when interpreting past treatment plant performance in order to predict future results. A large dataset is needed to cover a broad assessment. A shorter period of data review does not adequately cover the broad range of operational conditions that is typically to needed to evaluate performance. The TPS-95% values, relative to the average, also change with the data period. This means that periods of better or poorer performance may be present in one year and not another. The TPS-95% reflects sustainable periods to identify potential permit excursions. Since this level represents a monthly excursion every 1.7 years, a three-year dataset will potentially capture two months reflecting the TPS-95% value. A minimum of three years, and even longer is preferred to assess variability of the plant performance.

Nitrogen Performance

Figure 4-8 shows the results from various treatment plants removing nitrogen to low concentrations. The technologies used in these facilities vary. All of these except WSSC include tertiary filtration. Approximately two-thirds of the plants achieving TN below 3 mg/L (average) relies on an external carbon source to enhance denitrification (indicated by the * next to name). Only one of four plants above 3 mg/L average TN uses an external carbon source.

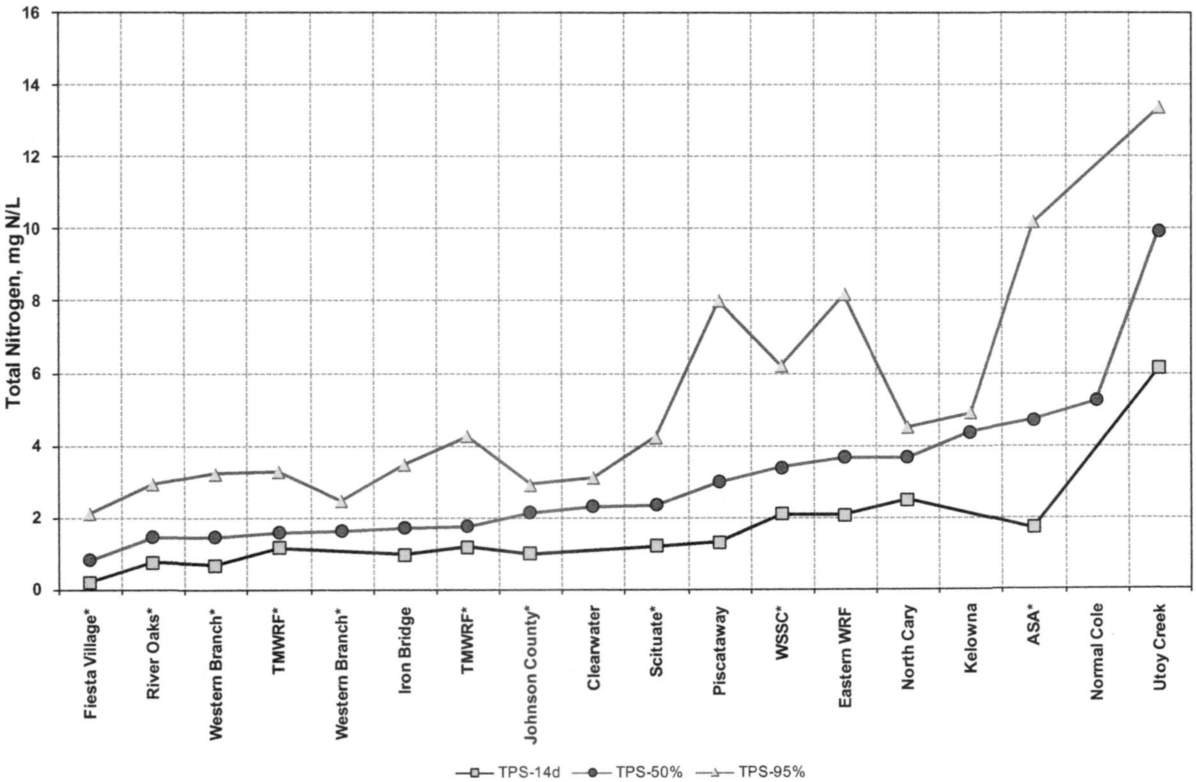

Figure 4-8. Total Nitrogen Reported by Treatment Plants.
(* indicate that external carbon is added, most using methanol)

Figure 4-9 shows EDON (effluent dissolved organic nitrogen) data from various municipal treatment plants reported by Stensel et al., 2009. The median EDON concentration is 1.2 mg/L, with concentrations ranging from 0.1 to 2.8 mg N/L. Treatment plants with a significant industrial contribution can observe much higher concentrations of EDON.

Lower effluent nitrogen can generally be achieved by using polishing or tertiary processes such as denitrification filters with methanol addition or membranes to remove essentially all particulate nitrogen. These technologies add both construction and operational costs. Treating recycle streams with chemicals to coagulate and remove colloidal particles has shown some success in reducing EDON.

EDON cannot be removed from effluent with current technologies (barring RO). New technologies may be required to change the nature of the EDON and make it treatable. While inorganic nitrogen can theoretically be reduced to low concentrations (below 0.5 mg/L) removing EDON is very difficult.

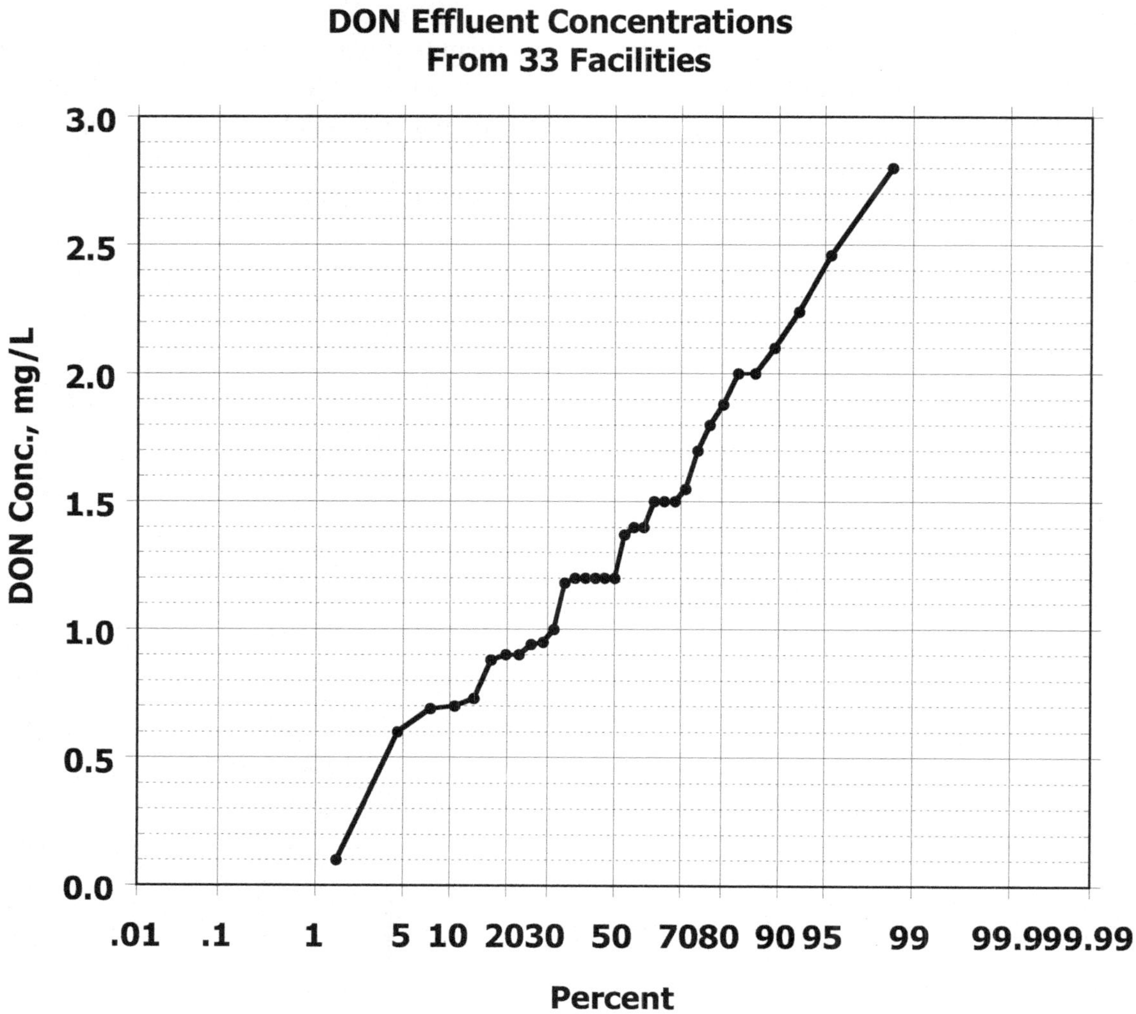

Figure 4-9. EDON Measured in Treatment Plants.

Table 4-7 shows the variability in performance for the nitrogen removal facilities in terms of the ratio between the best and average performance (3.84th / 50th percentile) and also the reliable performance (50th / 95th percentile). These two ratios indicate how much the performance deviates from the average performance to both lower and upper levels.

The best performance is typically 40-70% of the average value. This indicates that the plant is operating reasonably close to the best performance level. Since most of the high performing facilities add an external carbon source, the chemical dose can be increased to improve nitrate removal. The reliable performance is approximately 1.5 to 2.5 times higher than the average performance.

The cause for the variable performance has not been established. It could be attributed to external factors (influent variability, cold weather, construction, etc.) or internal factors (chemical dose, operator error, mechanical breakdown, chemical feed supply, etc.). This variability in performance needs to be accounted for when designing a process to provide the desired performance and reliability.

Table 4-7. Summary of Nitrogen TPS Values and Variability.

	TN - TPS-14d	TN - TPS-50%	TN - TPS-95%	3.84/50th	95/50th
Fiesta Village*	0.21	0.83	2.11	0.26	2.54
River Oaks*	0.78	1.45	2.92	0.54	2.01
Western Branch*	0.66	1.47	3.2	0.45	2.18
TMWRF*	1.17	1.59	3.27	0.74	2.06
Western Branch*		1.63	2.46		1.51
Iron Bridge	0.98	1.72	3.46	0.57	2.01
TMWRF*	1.1995	1.774	4.257	0.68	2.40
Johnson County*	1	2.14	2.9	0.47	1.36
Clearwater		2.32	3.1		1.34
Scituate*	1.21	2.37	4.22	0.51	1.78
Piscataway	1.3	3	8	0.43	2.67
WSSC*	2.1	3.4	6.2	0.62	1.82
Eastern WRF	2.09	3.67	8.18	0.57	2.23
North Cary	2.5	3.67	4.5	0.68	1.23
Kelowna		4.38	4.9		1.12
ASA*	1.72	4.72	10.18	0.36	2.16
Normal Cole		5.25			
Utoy Creek	6.14	9.94	13.37	0.62	1.35

Phosphorus Performance

Figure 4-10 show the results from various treatment plants removing phosphorus to low concentrations. The technologies used in these facilities vary. Many of the best performing facilities (effluent TP less than 50 µg/L) are small (less than 3 mgd capacity) and do not include anaerobic digestion. Some include aerobic digesters. The tertiary treatment at these best performing facilities typically includes two stage solids removal with clarification and conventional filtration or better final barrier to remove "all" particulate species.

Nearly two-thirds of the facilities show better than 100 µg/L effluent TP at a median level. The TPS-95% value, reduces the number facilities performing to this level to about a 1/3rd. Note that the performance graph is on a **log scale** and TPS-95% values below 200 µg/L are only achieved by about half of the facilities.

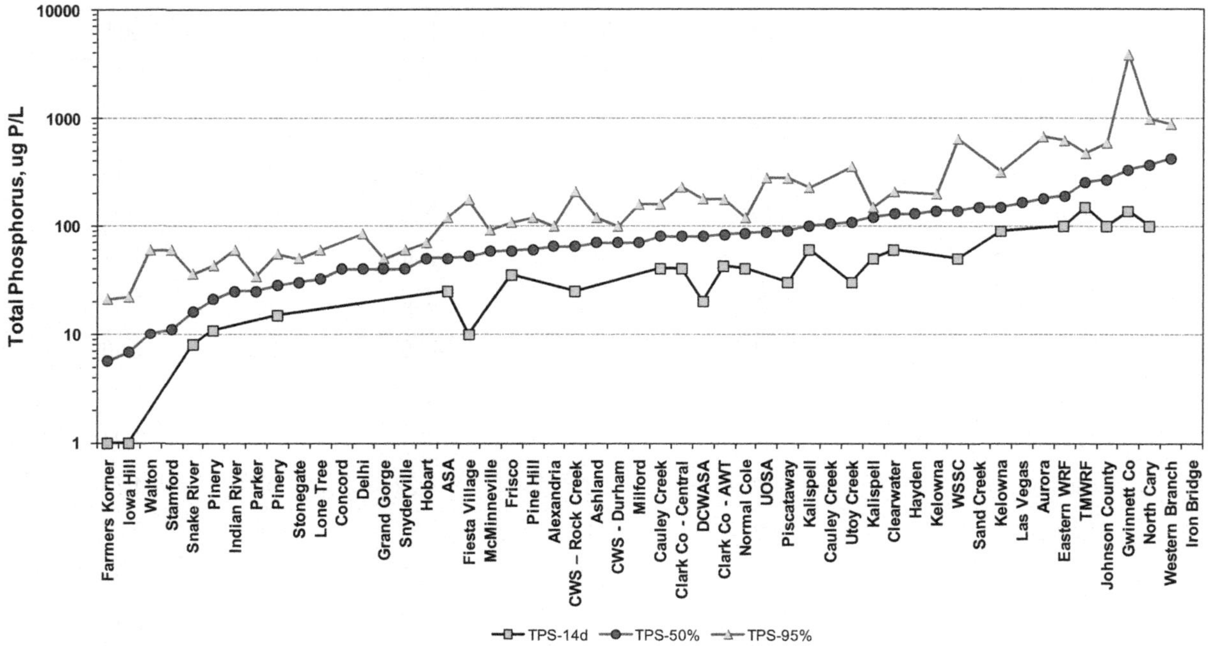

Figure 4-10. Total Phosphorus Reported by Treatment Plants.

Achieving low effluent phosphorus requires complete removal of particulate phosphorus species. This can be achieved with excellent solids separation processes such as filters and membranes. It also requires careful control over chemical dosing and provisions for pH adjustment, coagulation and flocculation to enhance particle removal.

EDNRP (effluent dissolved non-reactive phosphorus) cannot be removed from effluent with current technologies (barring possibly RO). New technologies may be required to change the nature of the EDNRP and make it treatable. While inorganic and particulate phosphorus can theoretically be reduced to low concentrations (below 5 μg/L in some cases), removing EDNRP is very difficult. Figure 4-11 shows results from a pilot study where 10-15 μg/L DNRP resisted treatment with four different technologies. In each case, the residual DNRP remained remarkable similar, indicating the difficulty to remove this with chemical addition and solids separation.

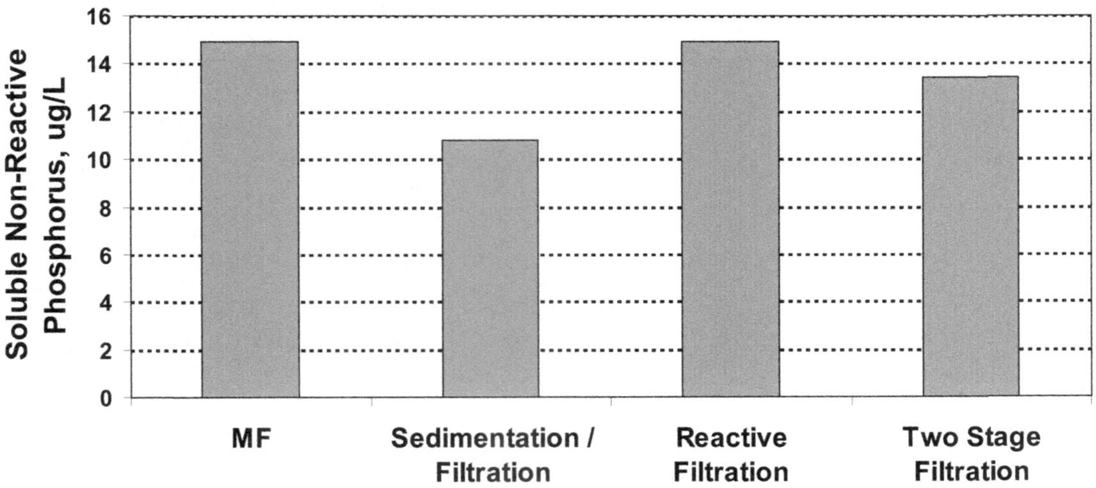

Figure 4-11. Phosphorus Species Remaining Following Four Different Chemical Treatment Options Using Alum and Ferric. Results show the persistence of dissolved non-reactive phosphorus in all processes (data taken from Benisch et al., 2007).

Table 4-8 shows the variability in performance for the phosphorus removal facilities in terms of the ratio between the best and average performance (3.84th / 50th percentile) and also the reliable performance (50th/95th percentile). These two ratios indicate how much the performance deviates from the average performance to both lower and upper levels.

The best performance is typically 40-60% of the average value. However, a few plants have very low values for low performance, less than 20% of the average performance. The reliable performance is highly variable and ranges from approximately 1.5 to 4 times higher than the average performance. The reason for the high variability is uncertain.

The cause for the variable performance has not been established. It could be attributed to external factors (influent variability, cold weather, construction, etc.) or internal factors (chemical dose, operator error, mechanical breakdown, chemical feed supply, etc). This variability in performance needs to be accounted for when designing a process to provide the desired performance and reliability.

Table 4-8. Summary of Phosphorus TPS Values and Variability (ug/l).

Plant	TP - TPS-14d	TP - TPS-50%	TP - TPS-95%	3.84/50th	95/50th
Farmers Korner	1	5.6	21	0.18	3.75
Iowa Hill	1	6.7	22	0.15	3.28
Walton		10	60		6.00
Stamford		11	60		5.45
Snake River	8	16	36	0.50	2.25
Pinery	11	21	43	0.52	2.05
Indian River		25	60		2.40
Parker		25	34		1.36
Pinery	15	28	55	0.54	1.96
Stonegate		30	50		1.67
Lone Tree		32	60		1.88
Concord		40			
Delhi		40	85		2.13
Grand Gorge		40	50		1.25
Snyderville		40	60		1.50
Hobart		50	70		1.40
ASA	25	50	120	0.50	2.40
Fiesta Village	10	52	176	0.19	3.38
McMinneville		58	92		1.59
Frisco	35	59	108.5	0.59	1.84
Pine Hill		60	120		2.00
Alexandria		65	100		1.54
CWS – Rock Creek	25	65	210	0.38	3.23
Ashland		70	120		1.71
CWS - Durham		70	100		1.43
Milford		70	160		2.29
Cauley Creek	40	80	160	0.50	2.00
Clark Co - Central	40	80	231	0.50	2.89
DCWASA	20	80	180	0.25	2.25
Clark Co - AWT	43	83	177	0.52	2.13
Normal Cole	40	86	120	0.47	1.40
UOSA		88	282		3.20
Piscataway	30	90	280	0.33	3.11
Kalispell	60	100	230	0.60	2.30
Cauley Creek		105			
Utoy Creek	30	110	360	0.27	3.27

Plant	TP - TPS-14d	TP - TPS-50%	TP - TPS-95%	3.84/50th	95/50th
Kalispell	50	121	150	0.41	1.24
Clearwater	60	130	210	0.46	1.62
Hayden		130			
Kelowna		139	200		1.44
WSSC	50	140	650	0.36	4.64
Sand Creek		150			
Kelowna	90	150	320	0.60	2.13
Las Vegas		166			
Aurora		180	685		3.81
Eastern WRF	100	190	630	0.53	3.32
TMWRF	150	260	480	0.58	1.85

Simultaneous Nitrogen and Phosphorus Removal

Some facilities can achieve both low nitrogen and phosphorus concentrations. Figure 4-12 shows the TPS-50% values for 10 facilities. It is interesting to note that the nitrogen and phosphorus removal performance moves in opposite directions: as nitrogen removal improves, phosphorus removal decreases fairly consistently in the figure.

The optimal balance and ability to reduce both nitrogen and phosphorus remains an open question. The challenges with simultaneous N & P removal relates to providing phosphorus during effluent nitrogen polishing (denitrification) to sustain growth. At the same time, phosphorus removal requires complete particulate phosphorus removal and also avoiding phosphorus release from the solids during the solids separation process. These processes most likely require separate treatment.

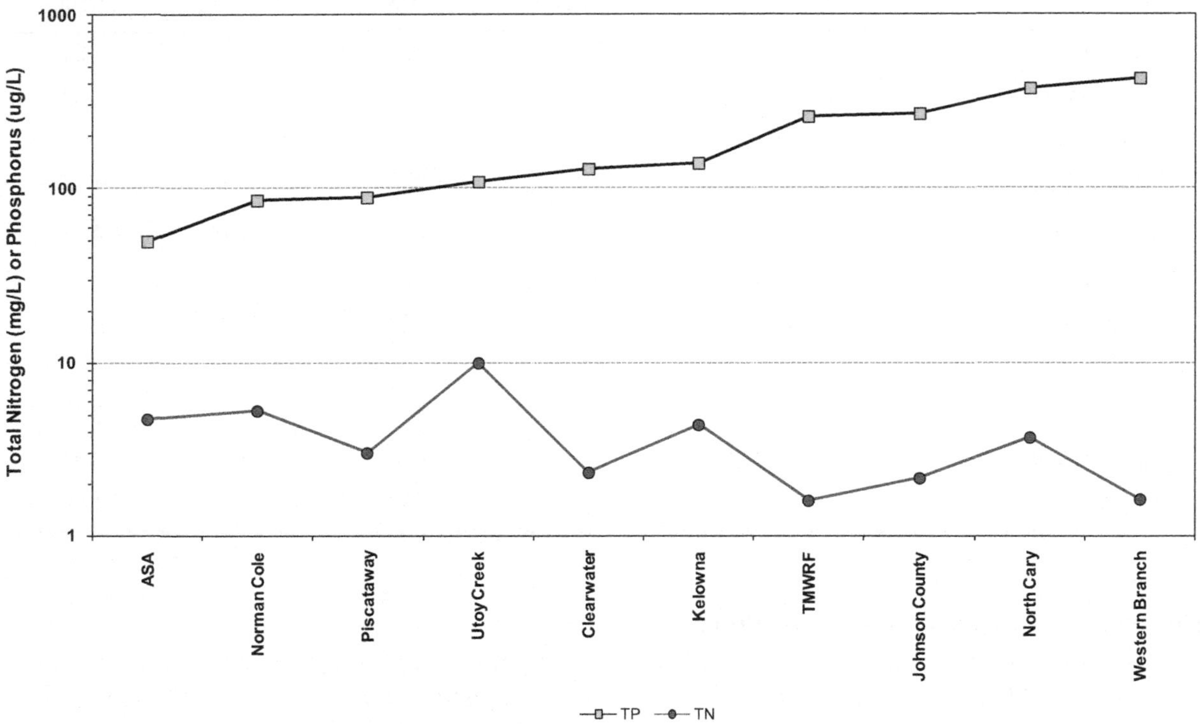

Figure 4-12. Treatment Plants Achieving Both Low Total Nitrogen and Total Phosphorus (TPS-50%).

4.4 Summary

- It is important to know the different nutrient species in wastewater influent and effluent in order to remove nutrient to low concentrations. Nutrient species can be divided into dissolved or particulate fractions, reactive or nonreactive fractions, and biodegradable/bioavailable or refractory species.
- Particulate species can be reduced using effective filtration processes. Dissolved species are removed through biological or chemical reactions. Inorganic species (ammonia, nitrate, phosphate, etc) are amenable to treatment and can be effectively removed. Some complex, often organic, compounds appear more difficult to remove.
- A fraction of dissolved nitrogen and phosphorus species remains in the effluent following treatment. This refractory species resist conventional treatment processes. The nature and abundance of refractory nitrogen and phosphorus species is not known. Standard methods for measuring the species the refractory species are not available. These refractory species are suspected of being organic compounds.
- Many technologies are available to remove nitrogen and phosphorus with new technologies emerging. Some processes will remove a specific nutrient while other can remove multiple nutrients. The performance of processes becomes linked. For example, since biological denitrification and biological phosphorus removal processes compete for readily biodegradable organic substrate it becomes challenging to design and operate a system with both processes. Another example is when phosphorus removal leads to a phosphorus deficiency in a tertiary denitrification process.

- Operational data from full scale facilities demonstrate the variability seen in plants. External and internal factors impacts treatment performance. External factors include variability in wastewater composition from day to day, weather (cold, rain), industrial discharges, changes in the service area, and many others. Internal factors include operator input, construction activities, equipment failure, chemical supply, control system failure, maintenance requirements (taking basins off-line, cleaning), internal recycle streams, and many others.
- Variable loads and operating conditions are unavoidable in wastewater treatment, with slug or peak loads negatively impacts treatment performance. Peak loads from internal processes (such as solids processing recycle streams) and external processes (industry, peak loads, etc) must be mitigated as much as possible to maximize treatment performance.
- Technology Performance Statistics (TPS) should be used to assess full scale plant performance data. The TPS addresses the variation in performance that must be accounted for when designing, operating, or predicting the capability of the process. The best achievable performance of a process (the 14-day performance or TPS-14d) is significantly lower than the average performance – typically 40-60% of the average performance and in some case as low as 20% of the average performance. The reliable achievable performance (the 95th value or TP-95%) can be 1.5 to 4 times the average performance. The reason for this large range is not clear. This means that reliable performance of a process is 5 to 10 times the best achievable performance.
- Permit structures that incorporate treatment and performance variability will provide an avenue to avoid overdesigning facilities to accommodate the worst case scenario under all operating conditions.

4.5 References

Benisch, M., Clark, D., Neethling, J.B., Fredrickson, H.S., Gu, A. 2007. *Tertiary Phosphorus Removal Pilot Tests Technology Limits In Coeur D'Alene, ID*. Water Practice, 1(13), 1-13.

EPA. 2008. *Municipal Nutrient Removal Technologies Reference Document*, EPA 832-R-08-006, September 2008. http://www.epa.gov/OWM/mtb/publications.htm

Esvelt, L.A. and Esvelt, M.H. 2006. *Advanced P Removal - Colorado Area Wastewater Treatment Plants*, Report by Esvelt Environmental Engineering, Spokane, WA.

Khan, E. 2007. Development of Technology Based Biodegradable Dissolved Organic Nitrogen (BDON) Protocol *Presentation at Water Environment Research Foundation and Chesapeake Bay Science and Technology Advisory Committee Workshop*, Baltimore, MD, September 27/28, 2007.

Lancaster, C.D. and Madden, J. 2008. Not so Fast! *Impact of Recalcitrant Phosphorus Fractions on the Ability to Meet Lowering Phosphorus Limits*. Conference Proceedings, WEFTEC08, Chicago, IL.

Neethling, J.B., Bakke, B., Benisch, M., Gu, A.Z., Stephens, H., Stensel, H.D., and Moore, R. 2005. *Factors Influencing the Reliability of Enhanced Biological Phosphorus Removal.* Project #01-CTS-3. Water Environment Research Foundation. Alexandria, VA.

Neethling, JB Neethling, Mario Benisch, Dave Clark, April Gu. *Phosphorus speciation provides direction to produce 10 μg/L*, IWA WEF 2007 Nutrient Conference, Baltimore, MD.

Neethling, J.B., Stensel, H.D., Parker, D., Bott, C., Murthy, S., Pramanik, A., and Clark, D. 2009. What is the Limit of Technology (LOT)? A Rational and Quantitative Approach. WEF/IAW conference Nutrient Removal 2009, Washington, D.C. June 28-July 1, 2009.

Pagilla, K. 2007. *Presentation at Water Environment Research Foundation and Chesapeake Bay Science and Technology Advisory Committee Workshop*, Baltimore, MD, September 27/28, 2007.

Parkin, G. and P.L. McCarty. 1981. Sources of soluble organic nitrogen in activated sludge effluents. *Journal Water Pollution Control Federation*. January 1981.

Ragsdale, D. 2007. *Advanced Wastewater Treatment to Achieve Low Concentration of Phosphorus* (Region 10, April 2007) http://yosemite.epa.gov/r10/water.nsf/Water+Quality+Standards/AWT-Phosphorus

Reynolds, D. and Clark, D. 2005. *Evaluation of Exemplary WWTPs Practicing High Removal of Phosphorus* (Dave Reynolds, CH2MHILL and Dave Clark, HDR). Study for Spokane River Dischargers, November 21, 2005.

Stensel, H.D., et.al. 2008. *Dissolved Organic Nitrogen (DON) in Biological Nutrient Removal Wastewater Treatment Processes*. Water Environment Research Foundation, November 2008. http://www.werf.org/nutrients/LOTDissolvedOrganicNitrogen.

WEF. 2009. *Design of Municipal Wastewater Treatment Plants, Manual of Practice No. 8,* Fifth Edition. Water Environment Federation, Alexandria, VA, 2008.

WEF/WERF. 2008. "Demonstrated Processes for Limit of Technology Nutrient Removal – Achievable Limits and Statistical Reliability." C. Bott, D. Parker, JB Neethling (chair). Preconference Workshop 101, WEFTEC 2008, Water Environment Federation Technical Exhibition and Conference, Chicago, IL.

WEF/WERF. 2009. Real World LOT Nutrient Removal: Achievable Limits and Statistical Reliability. C. Bott and D. Parker (chair). Preconference Workshop 216, WEFTEC 2009, Water Environment Federation Technical Exhibition and Conference, Orlando, FL.

CHAPTER 5.0

EFFLUENT DISCHARGE PERMITTING

Surface water nutrient discharges should receive special consideration in discharge permitting. Unlike BOD, ammonia nitrogen, and some toxic pollutants that can have acute effects in the aquatic environment, total nitrogen and phosphorus have seasonal impacts on receiving waters. Therefore, distinction should be made from these other effluent parameters upon which much of the existing EPA permit writer's guidance is based. Appropriate NPDES discharge permit structures for nutrients should be based on long averaging periods linked to the specific waterbody response to nutrient enrichment, such as seasonal limits based on long-term average values, or total loading for the compliance period (e.g., total pounds discharged on an annual or seasonal basis).

It is also important that consideration be given to variability and reliability of effluent performance from advanced nutrient removal facilities, especially those operating at low or very low levels. Appropriate NPDES permitting methodologies will avoid compliance issues that are immaterial to surface water quality protection. Short-term limitations, such as maximum daily and maximum weekly, should not be imposed for nutrients. Also, overly conservative assumptions as the bases for limit derivation, such as restrictive low frequency of occurrence receiving water flows (e.g., 7-day $Q_{10)}$ and extreme and improbable coincident events, such as statistical extremes occurring in both receiving waters and effluent discharge quality, should be avoided.

This chapter presents a discussion of nutrient discharge permitting issues and some of the special considerations associated with appropriate limits for nitrogen and phosphorus. Chapter 6.0 presents example discharge permits with nutrient limits as references.

5.1 Nutrient Discharge Permit Structures

5.1.1 Typical Permit Writer's Guidance

The appropriate averaging period for nutrient discharges depends on the sensitivity of the waterbody to water quality degradation and where the discharge is in the watershed. EPA's *NPDES Permit Writer's Manual* (EPA, 1996) states that for municipal wastewater treatment plants, permit limits should be expressed in average monthly and average weekly limits. Maximum daily limits can be used for toxics in order to capture acute toxicity criteria. In general, averaging periods for nutrient discharges can be longer due to slower responses between discharge and water quality degradation. For larger water bodies, such as bays, sounds, estuaries, and lakes, a monthly or yearly averaging period is more appropriate. In some cases, weekly average nutrient discharges are appropriate. Daily discharges are rarely appropriate given the lack of response in degraded water quality over the course of a single day for nutrient discharges.

However, daily mass and concentration limits are being imposed in some new NPDES permits, especially for industrial facilities.

5.2 Challenges in Discharge Permit Structures

5.2.1 Translation of TMDL Requirements to Effluent Discharge Permits

The TMDL sets the allowable pollutant load to a water body and allocates this load among the various point and non-point source dischargers. The permit writer then translates the wasteload allocation, as appropriate, into discharge permits. The NPDES permitting regulations at 40 C.F.R. Sec. 122.44(d) to describe how TMDL allocations are used in permitting:

> *"122.44 (d) Water quality standards and State requirements: any requirements in addition to or more stringent than promulgated effluent limitations guidelines or standards under sections 301, 304, 306, 307, 318 and 405 of CWA necessary to:....."*
>
> *(vii) When developing water quality based effluent limits under this paragraph the permitting authority shall ensure that:...."*
>
>> *(A) The level of water quality to be achieved by limits on point sources established under this paragraph is derived from, and complies with all applicable water quality standards; and*
>> *(B) Effluent limits developed to protect a narrative water quality criterion, a numeric water quality criterion, or both, are consistent with the assumptions and requirements of any available wasteload allocation for the discharge prepared by the State and approved by EPA pursuant to 40 CFR 130.7."*

EPA's *NPDES Permit Writer's Manual* (EPA, 1996) also addresses TMDLs and wasteload allocations.

Water quality (TMDL) and permitting (NPDES) programs are administered by separate staff groups within state regulatory agencies. Communication about the intent of the TMDL and the specifics required for the preparation of an NPDES permit then becomes very critical. The permitting authority is responsible for interpreting the water quality standards and TMDLs to develop the effluent limitations for the discharge. Their responsibility includes providing sufficient documentation in the administrative record to show how the NPDES permit was developed and how the compliance requirements will achieve the water quality standards and TMDL. Additionally, the permitting authority is responsible for the periodic assessment of water quality conditions based on the discharger and state monitoring data, and is responsible for determining if the permit is appropriate for meeting the water quality standards. Roles and responsibilities of the various surface water programs are summarized as follows, depending upon the individual state or organization:

The Permitting authority is responsible for reviewing the Permit Application and issuing a permit to the facility with limits that are protective of water quality standards. These permits will include Monitoring and Reporting requirements.

The Enforcement responsibility for the permit limits may, or may not, be contained in the same group as those responsible for writing the permits. The enforcement staff are responsible for ensuring that the facility is in compliance with the permit limits and review the Discharge

Monitoring Reports for exceedances. The assumption is that if the discharger is meeting permit limits, that they are being protective of water quality standards.

There are separate responsibilities for TMDL development and monitoring assessment of the in stream water quality. That is not the responsibility of the permitting authority.

Since NPDES permit writers may not involved with the development of TMDLs, there is the potential for a lack of understanding of the critical water quality issues within the watershed or the intentions of the TMDL. The permit writer may have a different perspective on the water quality standards and the receiving water requirements than the TMDL authors. Often, draft NPDES permits are based on pre-formulated guidance for permit structures, including monthly, weekly, and daily effluent limits that may not necessarily be appropriate for the situations involving nutrients. The watershed response to nutrient enrichment is generally over a seasonal time period longer than monthly or weekly time frames commonly used as the basis for NDPES permit limitations. Maximum weekly and maximum daily effluent limits for nutrients are overly restrictive and unnecessary to protect water quality from nutrient effects. However, NPDES permit guidance based on control of toxics results in a common permit structure incorporating maximum weekly and maximum daily effluent limits. While effluent restrictions over short time periods are necessary to protect receiving water quality from the discharges of toxics, such as ammonia nitrogen, chlorine residuals, and metals, it is seldom necessary for control of total nitrogen and phosphorus discharges.

The permitting authority may use all available tools to translate TMDLs and their wasteload allocations into enforceable effluent limitations in discharge permits. "For example, while the NPDES permitting regulations require 'daily maximum' limits for continuous discharges from some point sources, the same regulations specifically authorize 'average weekly' and 'average monthly' limitations, for discharges from publicly owned water treatment plants. Moreover, the regulations further authorize the permit writer to use other unspecified units of time if it is impracticable to calculate daily, weekly or monthly limitations. For non-continuous discharges, the regulations provide flexibility as to the manner in which such discharges are to be limited based on a consideration of factors, including frequency, total mass, maximum rate of discharge of pollutants and prohibition or limitation of specified pollutants by mass, concentration or other appropriate measure" (EPA, 2006a).

The wasteload allocation is expressed in the TMDL and the permit writer must translate the TMDL into compliance requirements that are appropriate to limit the discharge of the pollutant and protect the receiving waterbody.

5.2.2 Appropriate Averaging Periods for Nutrient Limits

Surface water nutrient discharges require special considerations to distinguish them from other effluent parameters, in particular toxic parameters, upon which much of the existing EPA permit writer's guidance is based. Appropriate NPDES discharge permit structures for nutrients should include long averaging periods, such as annual or seasonal limits based on total loading over long periods or annual or seasonal averages. Consideration should be given to variability and reliability in both the receiving waters and in the effluent performance from wastewater treatment systems.

The NPDES regulations (40 CFR 122.45(d)) require that all permit limits be expressed as average monthly limits and average weekly limits for publicly owned treatment works (POTWs) and as both average monthly limits and maximum daily limits for all others, unless "impracticable." EPA established the basis for appropriate effluent limits for Chesapeake Bay in a 2004 memorandum that combines considerations of water quality responses with scientific and policy analysis (Hanlon, 2004). EPA found that NPDES effluent limits for nitrogen and phosphorus expressed as an annual limit in lieu of daily maximum, weekly average, or monthly average are appropriate for protection of Chesapeake Bay and its tidal tributaries. EPA stated:

> *"...permit limits expressed as an annual limit are appropriate and that it is reasonable in this case to conclude that it is "impracticable" express permit effluent limitations as daily maximum, weekly average, or monthly average effluent limitations."*

EPA found that establishing appropriate permit limits for nitrogen and phosphorus to protect water quality in Chesapeake Bay was different from setting limits for other parameters, such as toxics for the following reasons:

- The exposure period of concern for nutrients loadings to Chesapeake Bay and its tidal tributaries is very long
- The area of concern is far-field (as opposed to the immediate vicinity of the discharge
- The average pollutant load rather than the maximum pollutant load is of concern.

The 2004 EPA memorandum notes that applicability to smaller scale embayments and tributaries was not considered. It is also noted that the annual average approach does not apply to other parameters that may impact dissolved oxygen such as BOD and ammonia nitrogen.

Unlike toxics and conventional parameters that have a direct and immediate impact on water quality, nutrients have no direct or immediate impact and must be processed in the aquatic environment in order to have an impact. Nutrient assimilation and processing delays and buffers the time between discharge and the receiving water effect. The 2004 EPA memorandum further distinguishes appropriate nutrient permit limits from the guidance provided in EPA's "Technical Support Document for Water Quality Based Toxics Control" (TSD) (EPA, 1991). The TSD statistical procedures for acute and chronic aquatic life protection is not applicable for periods more than 30 days. The exposure period for nutrients is longer than one month and may be up to a few years, and the average exposure rather than the maximum is of concern.

The TSD provides guidance for establishing daily and monthly limits for human health protection based on long exposure periods. However, this is not appropriate for nutrients because it is based on a steady state approach that assumes the effluent load is constant. EPA notes that to establish appropriate weekly or monthly limits would require temperature prediction over time because of the effect of temperature on treatment efficiency. Because of the effect of temperature on treatment efficiency and the normal variation in ambient temperature over short periods of time, EPA found that for the Chesapeake Bay that it is impracticable to develop appropriate daily, weekly, or monthly limits for nutrients that are protective of a wasteload allocation expressed as an annual load.

5.2.3 Maximum Day and Maximum Week Dilemmas

Effluent discharge permit structures should avoid the creation of frameworks that result in compliance issues that are immaterial to surface water quality protection, such as maximum daily and maximum weekly limits, overly restrictive receiving water flow assumptions, and the assumption of extreme and improbable coincident events, such as statistical extremes in both receiving waters and effluent discharge quality.

Maximum weekly and maximum daily effluent limits for nutrients are overly restrictive and unnecessary to protect water quality from nutrient effects. Waterbody responses to nutrients occur over longer periods of time associated with the growth and decay of algae, eutrophication and hypoxia that may impair beneficial uses, deplete dissolved oxygen, or result in fish kills.

In a study of numerous lakes, researchers found that there was often a lag period of a few years in chlorophyll *a* response to changes in nutrient loading, but that there was correlation between chlorophyll *a* and nutrient concentrations on an annual basis (Jeppeson, 2005). As noted above, EPA determined that annual limits are appropriate for protection of Chesapeake Bay and its tidal tributaries (Hanlon, 2004). In Montana, Clark Fork River effluent discharge limits for nitrogen and phosphorus are based on 30Q10 low flows based on peripyton algae studies conducted to support the Voluntary Nutrient Reduction Program (VNRP) (Tri-State, 1998). Mechanistic modeling studies of nutrient enrichment on the Yellowstone River by the Montana Department of Environmental Quality have used a 15Q10 flow condition to define conditions associated with periphyton growth response based on a number of studies (Bothwell, 1993), (Horner,1990), (Walton, 1995), (Welch, 2007).

5.2.4 Effluent Mixing Zones

An effluent mixing zone is an area within a waterbody where a point source discharge undergoes initial dilution or mixing in the receiving water. Within the mixing zone, water quality standards may be exceeded as long as acutely toxic conditions are prevented and all beneficial uses, such as drinking water, fish habitat, recreation, and other uses are protected. In theory, the regulatory mixing zone allows for efficient natural pollutant assimilation. In practice, mixing zones can be used as long as the integrity of a water body is not impaired. Water quality standards must be met at the edge of a mixing zone.

The use of mixing zones and dilution appear to have questionable applicability to watershed impacts from nutrients since the effects of nutrients tend to be cumulative and caused by mass loadings rather than toxic effects associated with effluent concentration. Nevertheless, regulatory agencies may approach effluent permitting for nutrients using mixing zone concepts and regulations.

Mixing zones can be useful for discharges with effluent quality that does not meet water quality standards and where state regulations allow for additional effluent mixing in the receiving water. These mixing zones may be permitted for a variety of constituents, including metals, toxic compounds and temperature. The EPA has established mixing zone rules and allows states to adopt additional mixing zone regulations as part of the state's water quality standards.

There are multiple areas within mixing zones. These include the allocated impact zone (AIZ), legal mixing zone (LMZ), toxic dilution zone (TDZ), and the zone of initial dilution (ZID). Two typical areas include the acute and chronic mixing zone. The acute mixing zone is the area of initial dilution, sometimes referred to as the zone of initial dilution (ZID), where acute criteria are met at the edge of this zone. Beyond the acute mixing zone is the chronic mixing zone, a larger area where chronic criteria must be met.

Many states have used mixing zones for decades. The EPA issued its first guidance document on mixing zones in 1968, and reaffirmed their use in its 1993 Water Quality Standards Handbook. The EPA has numerous other guidance and reference documents on mixing zones, include the *Compilation of EPA Mixing Zone Documents* (EPA, 2006b). It should be noted that mixing zones are not always appropriate, as documented in the following excerpts from EPA guidance:

> *"...mixing zones that allow for elevated levels of bacteria in rivers and streams designated for primary contact recreation are inconsistent with the designated use and should not be permitted because these could result in a significant health risk." (EPA, 2008a)*
>
> *"EPA recommends that mixing zone characteristics be defined on a case-by-case basis after it has been determined that the assimilative capacity of the receiving system can safely accommodate the discharge. This assessment should take into consideration the physical, chemical, and biological characteristics of the discharge and the receiving system; the life history and behavior of organisms in the receiving system; and the desired uses of the waters. Mixing zones should not be permitted where they may endanger critical areas (e.g., drinking water supplies, recreational areas, breeding grounds, areas with sensitive biota). (EPA, 2007a)*

Before a mixing zone is granted, the completion of a mixing zone study is required. Mixing zone studies evaluate the effectiveness of the mixing of effluent and receiving water under a variety of conditions to ensure compliance with water quality standards.

Estimating dilution may be performed using mathematical modeling or through dye studies. Dye studies are permitted only to demonstrate and evaluate the mixing zone for existing discharges.

When a mixing zone is permitted, the state will still seek to keep the mixing zone as small as possible. The size of the area, or "zone," will depend upon how concentrated the wastewater discharge is, the water quality standards, the location of the discharge in relation to other features, and the flow or size of the waterbody. The physical dimensions of mixing zones vary according to the type of environment (marine, estuarine, freshwater), depth of discharge, available currents, and/or varying regulations concerning the allowable spatial dimensions.

5.2.4.1 Impaired Ambient Conditions

Impaired ambient water quality can create difficult situations for effluent discharge permitting since any additional contribution of nutrients may compound receiving water conditions and no cleaner water is available for dilution. By definition, impaired waterbodies that are 303(d) listed and require a TMDL may not have assimilative capacity to receive additional loadings. In some waterbodies, this has led to in-stream nitrogen and phosphorus target concentrations being applied at the end-of-pipe to effluent discharges. The result may be effluent limits that are below the limits of treatment technology.

5.2.5 Permit Requirements Beyond the Capability of Treatment Technology

The NPDES program requires that discharge permits include specific pollutant limitations. These discharge limits are initially set based on applicable treatment technology standards depending upon the specific pollutant or parameter, type of discharge or industry in the case of effluent guidelines. These technology-based limits are then evaluated to determine if the allowable discharges will comply with the receiving water quality requirements. If not, more restrictive limitations are to be established that are water quality-based. However, these water quality-based effluent limits (WQBELs) may represent levels that are beyond the capability of economically available treatment technology.

Dischargers facing these conditions must deal with treatment options that are very expensive to design, construct, operate, and maintain. Additionally, once operational the technologies are challenging to operate and maintain at such consistently low concentrations. With more dischargers competing for a smaller piece of the allowable pollutant allocation, requirements that exceed the capability of conventional and economical technology are becoming more common. This is the case with nitrogen and phosphorus limits with nutrient criteria continuing to be developed resulting in very low numeric standards.

There is not a common understanding or consensus between regulators, the public, and dischargers on the economics and feasibility of implementing such limitations that push the envelope of treatment technology capability. In fact, it is not yet clear if the available nutrient removal technologies are even able to consistently treat to such low concentrations, especially at higher flows.

When a TMDL requires that all available loadings are allocated to existing dischargers at the limits of treatment technology, any future needs for capacity expansion must be met by improvements in treatment technology or from diversion of effluent to land application or effluent reuse programs. An example of this situation is described in the South Carolina Department of Health and Environmental Control, *NPDES Permit Program Load Allocation*:

In situations where one or more existing dischargers are at their limits of treatment technology, the Bureau will reduce the permitted loadings of the dischargers that are not at the limits of treatment technology by the same percentage until the model predicts no water quality violations will occur. The new or expanded discharger will have the same percent reduction from their technology based limits. In this evaluation when an existing discharger reaches their limits of treatment technology, their loading will not be reduced any further and the modeling will be repeated using further reductions for the dischargers that have not reached their limits of treatment technology. If the evaluation reaches a point where all existing and proposed dischargers' loadings have been reduced to their limits of treatment technology and the model still predicts water quality violations will occur, the new or expanded discharger cannot be allowed as proposed. The new or expanded discharger may be allowed on a smaller scale than was originally proposed such that the total loading from all dischargers will meet water quality standards.

In situations where the Bureau determines that all existing dischargers are already reduced to their limits of treatment technology, the existing total loading to the stream cannot be reduced through better treatment. In this situation, the new or expanded discharger cannot be permitted to surface waters unless reductions are made in other ways. The Bureau will normally encourage existing dischargers to reduce their loadings by other means such as source reduction, recycling, land application of effluent, water conservation, alternate manufacturing processes, consolidation of facilities through regional planning, etc. In situations where the Bureau determines that the existing loading exceeds the allowed stream loading, the Bureau may require the actual loading to the stream to be reduced by the existing dischargers utilizing the above methods even when there is not a proposed new or expanding discharger.

These challenges arise when nutrient permit limits for a water quality based effluent limits (WQBELs) are lower than treatment technology can achieve. If there is no established trading program available in a state, a permittee may not be able to meet effluent limits when the WQBELs are written beyond what is achievable by current treatment technology in order to meet water quality standards. To determine whether trading may be allowable, it is important to recognize that only water quality based effluent limits are eligible for trading according to EPA's 2003 Trading Policy (EPA, 2003a):

> *"Technology-Based Trading. EPA does not support trading to comply with existing technology-based effluent limitations except as expressly authorized by federal regulations. Existing technology-based effluent guidelines for the iron and steel industry allow intraplant trading of conventional, nonconventional and toxic pollutants between outfalls under certain circumstances (40 CFR 420.03)."*
>
> *"Trading to Maintain Water Quality Standards. Trading may be used to maintain high water quality in waters where water quality standards are attained, such as by compensating for new or increased discharges of pollutants."*

If trading is not a feasible option and the discharger cannot meet permit limits set at or beyond the capabilities of treatment technology, then other alternatives must be considered. The discharger may have to store water during the low flow, or zero discharge period, or find an alternative means for disposal, such as land application or irrigation, water reclamation or reuse, or groundwater recharge.

5.3 Water Quality Off-Sets and Trading

5.3.1 Background

Water quality trading is an innovative approach to achieve water quality goals more efficiently. Trading is based on the fact that sources in a watershed can face very different costs to control the same pollutant. Trading programs allow dischargers facing higher pollution control costs to meet their regulatory obligations by purchasing environmentally equivalent (or superior) pollution reductions from another source at lower cost, thus achieving the same water quality improvement at lower overall cost.

The basis of trading is that a water quality goal is established and that sources within the watershed have significantly different costs to achieve comparable levels of pollution control. Water quality trading is a voluntary option that regulated point sources can use to meet their NPDES permit limits. In certain watersheds, trading can provide significant economic and environmental benefits. For example, the full scale implementation of trading could save $1 billion in wastewater treatment costs in the Chesapeake Bay (Chesapeake Bay Blue Ribbon Finance Panel Report, Oct. 2004) (EPA, 2008b).

Where watershed circumstances favor trading, it can be a powerful tool for achieving pollutant reductions faster and at a lower cost. Water quality trading will not work everywhere, however. Trading works best when:

- A "driver" motivates facilities to seek pollutant reductions, usually a TMDL or a more stringent water quality-based requirement in an NPDES permit;

- Sources within the watershed have significantly different costs to control the pollutant of concern;
- The necessary levels of pollutant reduction are not so large that all sources in the watershed must reduce as much as possible to achieve the total reduction needed – in this case there may not be enough surplus reductions to sell or purchase; and
- Watershed stakeholders and the state regulatory agency are willing to try an innovative approach and engage in trading design and implementation issues.

In January 2003, the EPA issued the *National Water Quality Trading Policy*, supporting the trading as an innovative and market-based approach to improving water quality. The EPA issued the *Water Quality Trading Assessment Handbook* in 2004, which provides guidance on how stakeholders can environmentally and economically determine whether trading is feasible in their watershed.

The EPA's 2007 publication *The Water Quality Trading Toolkit for Permit Writers* is the first "how-to" manual on designing and implementing water quality trading programs. The Toolkit helps NPDES permitting authorities incorporate trading provisions into permits (EPA, 2007b).

5.3.2 Application

The Water Environment Research Foundation (WERF) has conducted research that provides a common base for greater understanding of the approaches and programs that have reduced nutrient loads and improved water quality.

While technology continues to advance – and there are several secondary and tertiary treatment (advanced) technologies that offer great promise for the future – the team's research indicates that a "one size fits all" approach may negate current improvements as documented in several successful efforts led by some states and watersheds across the nation. Examining both the current state of the science on treatment technologies, as well as the market-based approaches to water quality goals, and applying these to varying hydrological and site-specific conditions should add to greater understanding and subsequent improvements in water quality.

Recent data indicate that non-point sources, such as agricultural/animal farming operations, automobile and industrial and power plant air emissions, and runoff pollution, are far larger sources of water quality impairments due to nutrients compared to point sources. Wastewater treatment facilities represent a smaller source of these nitrogen and phosphorus; however, they continue to play a vital role in improving and maintaining water quality in the nation's waterways, and can also help manage non-point sources. Various "tools" currently exist to help affected communities, including advanced wastewater treatment, source control, and trading.

5.3.3 Water Quality and Market-Based Approaches

A market-based approach to meet water quality goals represents successful innovation that is both efficient and cost-effective. Water-quality or watershed based trading allows dischargers to meet their regulatory or stakeholder obligations by using pollutant reductions created by another discharger within the watershed lower pollution control costs. It helps focus and provide resources and attention to areas with greater positive impact for water quality improvements.

The Long Island Sound Watershed was one of five watershed studies or demonstrations of trading projects funded by WERF nationwide that resulted in successful outcomes. This study, led by Moore, et al (2000), resulted in state legislation in Connecticut that allowed trading in the Long Island Sound watershed. Watershed permits were issued to nearly 80 utilities based on the work of this project, saving over $200 million in capital cost savings, while also meeting the state goal of reducing nitrogen loads by 70%, and establishing a Nitrogen Credit Exchange to administer the trading program.

The innovative Connecticut Water Quality Trading Program was awarded the First EPA "Blue Ribbon" Award in 2007 and serves as a successful model for other programs nationwide (EPA, 2008c). Additional information and resources on trading is available at the EPA Office of Water website (EPA, 2008d).

Valuable lessons learned from the five trading studies funded by WERF (Cherry Creek, CO (WERF 2000a); Long Island Sound, CT (WERF, 2000b); Kalamazoo River Basin (WERF, 2000c); Fox-Wolf Basin (WERF, 2000d); and Maryland (WERF, 2002)) have been synthesized in a recent book published by lead author Cy Jones (2006) for the wastewater community. It helps demystify trading by identifying and explaining the issues involved in making trading decisions and the various analyses the treatment facility needs to undertake. While solutions are unique to each watershed, the issues, incentives, requirements, and barriers to trading have common themes, and the book presents an excellent framework to address these needs. It identifies methods to solve trading problems and minimize risks, and provides useful information and examples of existing programs, designs, checklists, and resources.

WERF recently published a 32-page user guide (Project #02WSM1, 2007) and PowerPoint-based presentations on CD-ROM to help stakeholders interested in water quality credit trading fill gaps in their understanding of trading, augment their capabilities to undertake a trading program, and facilitate trading deliberations at the watershed-level to help more programs meet watershed goals and objectives. This CD-ROM includes a combination of information, analyses, guidance, examples, templates, checklists, model elements, and references that offer an organized, systematic process in the form of 11 PowerPoint presentations.

Market-based approaches have been successful in improving water quality in selected watersheds and are now ready for wider implementation. Wastewater treatment facilities can play a vital role in helping States and watersheds establish successful nutrient trading and credit exchange programs that also help non-point sources.

5.3.4 Novel NPDES Discharge Permits

Many dischargers have begun to approach NPDES permitting in creative ways to meet more stringent water quality criteria that test the limits of current treatment technology. Innovative options that dischargers have used to undertake NPDES permitting include:

- Watershed-based permitting
- Water quality offsets or water quality trading

In 2003, the EPA produced a brochure entitled *Watershed-based Permitting: Rethinking Permitting as Usual* (EPA, 2003). The brochure outlines various watershed-based permitting approaches and why they are beneficial. Four examples from across the country are described, and the EPA also recognizes the resources they offer in promoting watershed-based permitting, including stakeholder involvement efforts.

5.3.4.1 Case Studies

In northern Kentucky, the State legislation assigned the responsibility and ownership of most wastewater collection systems in a three-county area to Sanitation District No.1. Potential water quality degradation contributors include sanitary sewage, urban storm water runoff, rural storm water runoff and failing septic systems. The Sanitation District No. 1 completed the Watershed-Based Permitting Feasibility Assessment Report in March 2004, which includes steps for identification, the assessment and selection of watershed-based permitting, and the controls for one sub-watershed within the district's jurisdiction.

In the Lake Lewisville watershed in north Texas near the Dallas-Fort Worth Metroplex, City of Denton officials were proactive in approaching future NPDES requirements. Although Lake Lewisville did not have any 303(d) listed constituents, city officials saw the need to protect their water resources in the face of watershed urbanization. They began a program of watershed-scale monitoring, land use planning and management, and public awareness.

Through an EPA Section 319 grant project, the City of Denton is implementing a Watershed Protection Plan for one sub-watershed of Lake Lewisville. The plan's elements include:

- Identify the causes and sources of pollutants.
- Estimate load reductions.
- Describe the nonpoint source management measures that will need to be implemented.
- Estimate the amounts of technical and financial assistance needed.
- Describe the information and education component.
- Estimate a schedule for implementing the nonpoint source management measures.
- Describe the interim, measurable milestones.
- Develop a set of criteria that can be used to determine whether loading reductions are being achieved.
- Describe the monitoring component.

The City of Denton reported that the watershed-based approach requires additional labor hours and resources, commitment during the initial stages of planning and implementation, and a secure source of long-term funding, which can be challenging for municipalities to obtain. The city also notes that it is critical to involve the public early in a watershed-based approach. Denton was able to engage the public by packaging watershed-based information in an easy-to-read and understandable format and used the internet to disseminate information.

Through a cooperative agreement with the EPA, the Conservation Technology Information Center developed a guide for the agricultural industry explaining the benefits of water quality trading. The guide, "Getting Paid for Stewardship: An Agricultural Community Water Quality Trading Guide" (CTIC, 2006), explains key factors in the trading process. The guide also communicates that, besides benefiting the industrial or municipal facilities by helping them meet their regulatory requirements, the agricultural producers get paid for the trade.

An NPDES storm water permit issued in 2001 for the Cities of Milpitas, Palo Alto, and Santa Clara requires each permittee to develop and implement an Urban Runoff Management Plan.

5.3.5 Watershed-Based Permitting

Watershed-based NPDES permitting is a process that emphasizes addressing all stressors within a hydrologically-defined drainage basin, rather than addressing individual pollutant sources on a discharge-by-discharge basis. Watershed-based permitting can encompass a variety of activities, ranging from synchronizing permits within a basin to developing water quality-based effluent limits using a multiple discharger modeling analysis. The type of permitting activity will vary depending on the unique characteristics of the watershed and the sources of pollution impacting it. The ultimate goal of this effort is to develop and issue NPDES permits that better protect entire watersheds (EPA, 2008).

The 2007 Watershed-Based National Pollutant Discharge Elimination System (NPDES) Permitting Technical Guidance is a follow up to the 2003 Watershed-based National Pollutant Discharge Elimination System (NPDES) Permitting Implementation Guidance and provides greater detail concerning a number of permit development and issuance questions previously not addressed. The document is focused on helping NPDES authorities develop and issue NPDES permits that fit into an overall watershed planning and management approach with input from watershed stakeholders.

As defined in the Watershed Approach Framework (EPA, 1996), "[T]he watershed approach is a coordinating framework for environmental management that focuses public and private sector efforts to address the highest priority problems within hydrologically defined geographic areas, taking into consideration both ground and surface water flow." A watershed approach has three basic components, a geographic focus, sound management techniques based on strong science and data, and partnerships and stakeholder involvement (EPA, 2007c).

Implementing the NPDES program within a watershed framework can initially require additional time and effort on the part of the permitting authorities, the permittees, and other stakeholders. However, potential environmental and administrative benefits to this process exist. Developing comprehensive and simultaneous solutions to water quality problems, as well as setting priorities for implementing those solutions, should result in better and, potentially, faster water quality improvements for the resources invested.

5.3.5.1 Case Studies

In an effort to promote watershed-based permitting, EPA has published a series of case study fact sheets that describe various efforts throughout the country, some of which include nutrient trading. The EPA case studies include the following:

- General Permit for Nitrogen Dischargers in Long Island Sound: Final Permit
- Sand Creek Watershed, Colorado: Watershed-Based Selenium Standard
- Michigan Statewide Stormwater Permitting: Statewide Watershed-based MS4 Stormwater General Permit
- Tualatin River Watershed, Oregon: Clean Water Services Integrated Municipal Permit
- Rahr Malting Company: Final Permit

- Northern Kentucky Sanitation District No. 1: Permitting Approach
- Industrial Stormwater Discharges from Dredging at Marinas in Lake Tahoe
- Construction Stormwater Discharges from Land Disturbance in Lake Tahoe
- Waste Discharge Requirements for South Lake Tahoe, El Dorado County and Placer County
- Louisville and Jefferson County Metropolitan Sewer District (MSD)
- Neuse River Watershed, North Carolina: Neuse River Compliance Association Watershed-Based Permit
- Big Darby Creek Watershed, Ohio: Construction Watershed-Based General Permit
- Chesapeake Bay Watershed, Virginia: Watershed-based General Permit for Nutrient Discharges and Nutrient Trading
- Lake Lewisville Watershed, Texas: City of Denton Watershed Protection Program
- North Carolina Statewide Approach: Basinwide Planning and Permitting

5.3.5.2 General Permit for Nitrogen Dischargers in Long Island Sound: Final Permit

In the Long Island Sound, excessive nitrogen causes low dissolved oxygen concentrations in the western portion of Long Island Sound during the summer. Publicly-owned treatment works (POTWs) in Connecticut and New York are a dominant source of nitrogen. Through the Long Island Sound Study, a 2014 goal of 58.5% nitrogen reduction from baseline has been established for Connecticut and New York. Connecticut and New York have formalized the nitrogen reduction program in a TMDL approved by the EPA in April 2001. The entire State of Connecticut is within the Long Island Sound watershed (EPA, 2003).

Highlights of the permit include the following (EPA, 2003):

- Effective 1/1/2002
- 79 POTWs (dischargers of at least 20 pounds of total nitrogen (TN) per day)
- Annual end-of-pipe permit limits in pounds of TN per day for each POTW, apportioned by plant discharge volume to meet the aggregate state target.
- Facilities can purchase or sell nitrogen credits annually based on each facility's performance with respect to their annual limit.

5.3.5.3 Tualatin River Watershed, Oregon: Clean Water Services Integrated Municipal Permit

Clean Water Services (CWS) is a public utility (special services district) that operates four municipal wastewater treatment facilities, each with its own permit under the NPDES. CWS also has two industrial stormwater permits and is a co-permittee on a Municipal Separate Storm Sewer System (MS4) permit. The Tualatin River is the receiving stream for each of these permitted discharges. Oregon's Department of Environmental Quality (OR DEQ) issued total maximum daily loads (TMDLs) for the Tualatin River for ammonia, phosphorus, temperature, bacteria, and tributary dissolved oxygen (DO). In February 2004, OR DEQ issued a single watershed-based, integrated municipal permit to CWS. This permit incorporates the NPDES requirements for all four of CWS's advanced wastewater treatment facilities, its two industrial storm water permits, and its MS4 permit. A significant feature of the integrated permit is its inclusion of provisions for water quality credit trading involving temperature (thermal load), biochemical oxygen demand (BOD), and ammonia (EPA, 2007d).

5.3.5.4 Chesapeake Bay Watershed, Virginia: Watershed-Based General Permit for Nutrient Discharges and Nutrient Trading

In March 2003, the Chesapeake Bay Program (CBP) adopted new nutrient reduction goals as part of the Chesapeake 2000 Agreement. This Agreement was established to protect and restore water quality in the Chesapeake Bay by a January 1, 2011 deadline. The nutrient reduction goals established in this Agreement aim to decrease the amount of total nitrogen (TN) and total phosphorus (TP) entering the Bay by 110 million and 6.3 million pounds per year, respectively. Based on these target reductions for the Chesapeake Bay watershed, the CBP established nutrient load allocations for each of the eight tributary basins (i.e., sub-watersheds). Each state within the Chesapeake Bay drainage area then developed tributary strategies to achieve the nutrient reduction goals for each sub-watershed (EPA, 2007e).

The Virginia Department of Environmental Quality (DEQ), in conjunction with the Virginia Department of Conservation and Recreation (DCR) and the EPA, developed tributary strategies for the Virginia tributaries to the Chesapeake Bay. Each tributary strategy establishes total nutrient loading allocations for both point and nonpoint sources within each sub-watershed and outlines implementation plans to meet these allocations (EPA, 2007e).

Monitoring requirements for facilities covered under the General Permit.

Parameter	Monitoring Sample Type/Frequency		
	Design Flow		
	≥ 20.00 MGD	*1.00–19.99 MGD*	*0.04–0.99 MGD*
Total Nitrogen	24 HC*/ 3 Days Per Week	24 HC/ 1 Day Per Week	8 HC/ 2 Per Month (>7 Days Apart)
Total Phosphorus	24 HC/ 3 Days Per Week	24 HC/ 1 Day Per Week	8 HC/ 2 Per Month (>7 Days Apart)

*HC = hour composite (e.g. 24 HC = 24 hour composite sample)

Total monthly and year-to-date mass loads must be calc
lated as specified by the General Permit as follows:

$$\mathbf{ML = ML_{avg} * d}$$

ML = total monthly load (lbs/mo)

ML_{avg} = monthly average load as reported on discharge monitoring report (lbs/day)

d = number of discharge days in sampling month

$$\mathbf{AL - YTD = \Sigma_{(January - current\ month)}\ ML}$$

AL-YTD = calendar year-to-date annual load (lbs/yr)

ML = total monthly load (lbs/mo) as reported on discha
monitoring report

Figure 5-1. Figure from http://www.epa.gov/npdes/pubs/wq_casestudy_factsht13.pdf (U.S. EPA, 2007b).

Table 5-1. Watershed-Based Permit Case Study Details. All Watershed Permits Allow for Trading of Nutrient Credits.

Watershed	Permit Issued	Permit Controlled Nutrients	Limit Type (Mass or Concentration	Number of Plants	Industrial Stormwater
Long Island	1/1/2002	TN	Mass	79	
Tualatin River	2/26/2004	Temperature, bacteria, DO, ammonia, and phosphorus		4	2
Chesapeake Bay	1/1/2007	TN and TP	Mass	127*	

* Significant dischargers, defined as an existing facility that discharges 100,000 gallons or more per day, or an equivalent load, directly into tidal waters; or an existing facility that discharges 500,000 gallons or more per day, or an equivalent load, directly into nontidal waters.

5.4 References

Bothwell, M.L., D. Sherbot, A.C. Roberge, and R.J. Daley. 1993. Influence of Natural Ultraviolet-Radiation on Lotic Periphytic Diatom Community Growth, Biomass Accrual, and Species Composition – Short-Term Versus Long-Term Effects. Journal of Phycology. Vol. 29 Issue 1.

Chesapeake Bay. 2004. Saving a National Treasure: Financing the Cleanup of the Chesapeake Bay. http://www.chesapeakebay.net/content/publications/cbp_12881.pdf, http://www.chesapeakebay.net/publicationsearchresults.aspx?menuitem=14874

EPA. 1991. Technical Support Document for Water Quality Based Toxics Control. EPA 505/2-90-001, March 1991.

EPA. 1996. U.S. EPA NPDES Permit Writers' Manual, EPA-833-B-96-003, December 1996.

EPA. 2003. Water Quality Trading Policy. EPA Office of Water. January 13, 2003.

EPA. 2006a. Establishing TMDL "Daily" Loads in Light of the Decision by the U.S. Court of Appeals for the D.C. Circuit in Friends of the Earth, Inc. v. EPA, et al., No. 05-5015 (April 25, 2006) and Implications, for NPDES Permits, Nov 15, 2006 Memorandum.

EPA. 2006b. Compilation of EPA Mixing Zone Documents. EPA 823-R-06-003.

EPA. 2007a. Water Quality Standards Handbook: Second Edition. EPA-823-B-94-005. July 2, 2007.

EPA. 2007b. Water Quality Trading Toolkit for Permit Writers. Office of Wastewater Management Water Permits Division. EPA 833-R-07-004. August 2007. http://www.epa.gov/owow/watershed/trading/WQTToolkit.html

EPA. 2008a. "Initial Zones of Dilution for Bacteria in Rivers and Streams Designated for Primary Contact Recreation," Memorandum from Ephraim King, EPA Office of Science and Technology, November 14, 2008.

EPA. 2008b. http://www.epa.gov/owow/watershed/trading.htm.

Hanlon, James H., Director Office of Wastewater Management, Memorandum to Jon Capacasa, Director Water Permits Division, EPA Region and Rebecca Hammer, Director Chesapeake Bay Program Office, "Annual Permit Limits for Nitrogen and Phosphorus for Permits Designed to Protect Chesapeake Bay and its tidal tributaries from Excess Nutrient Loadings under the National Pollutant Discharge Elimination System," March 3, 2004.

Horner, R.R., E.B Welch, M.R. Seeley, and J.M. Jacoby. 1990. Responses of Periphyton to Changes in Current Velocity, Suspended Sediment and Phosphorus Concentration. Freshwater Biology. Vol. 24 Issue 2.

Jeppeson et al. 2005. Lake responses to reduced nutrient loading—an analysis of contemporary longterm data from 35 case studies. *Freshwater Biology,* 50: 1747–1771.

Walton, S.P., E.E. Welch, and R.R. Horner. 1995. Stream Periphyton Response to Grazing and Changes in Phosphorus Concentration. Hydrobiologia. Vol. 302 Issue 1.

Water Environment Research Foundation. 2008. Accessed: October 1, 2008. www.werfnutrientchallenge.org.

Welch, E.B. and J.M. Jacoby. 2007. Pollutant Effects in Fresh Waters: Applied Limnology. Taylor & Francis. New York, NY.

Tri-State Implementation Council. 1998. Nutrient Target Subcommittee, Clark Fork River Voluntary Nutrient Reduction Program (VNRP), August 1998. http://www.tristatecouncil.org/documents/98vnrp_doc.pdf

Water Quality and Market-Based Approaches:

EPA. 2008c. http://www.epa.gov/ow/waternews/2007/071113.html#1.

EPA. 2008d. http://www.epa.gov/owow/watershed/trading.htm.

Jones, C., Bacon, L., Kieser, M.S., Sheridan, D., Water-Quality Trading. 2006. A Guide for the Wastewater Community, McGraw-Hill / WEF / WERF.

Moore, R.E., Overton, M., Norwood, R.J., DeRose, D. 2000. Nitrogen Credit Trading in the Long Island Sound Watershed, WERF / WEF (97-IRM-5B)

Reckhow, K.H., et al. 2006. Water Quality Indicators: Nutrient Impacts on Chlorophyll or Algae Species Composition, WERF / IWAP. (02ECO1).

Warren-Hicks, W.J., Parkhurst, B.R., Bartell, S.M., and Smart, M.M. 2005. Technical Approaches for Setting Site-Specific Nutrient Criteria, WERF / WEF / IWAP (99WSM3).

Water Environment Research Foundation. 2000a. Phosphorus Credit Trading in the Cherry Creek Basin: An Innovative Approach to Achieving Water Quality Benefits, Project #97-IRM-5A. Alexandria, VA.

Water Environment Research Foundation. 2000b. *Nitrogen Trading in the Long Island Sound Watershed,* Project #97-IRM-5B. Alexandria, VA.

Water Environment Research Foundation, 2000c, *Phosphorus Credit Trading in the Kalamazoo River Basin: Forging Nontraditional Partnerships*, Project #97-IRM-5C. Alexandria, VA.

Water Environment Research Foundation. 2000d. *Phosphorus Credit Trading in the Fox-Wolf Basin: Exploring Legal, Economic, and Technical Issues*, Project #97-IRM-5D. Alexandria, VA.

Water Environment Research Foundation. 2002. Nitrogen Trading in Maryland: A Market Analysis for Establishing a Statewide Framework, Project #97-IRM-5E. Alexandria, VA.

Water Environment Research Foundation, 2007. *Water Quality Credit Trading: Powerpoint Workbook for a Detailed Assessment of Opportunities and Options*. Project #02-WSM-1. Alexandria, VA.

NPDES Permitting:

CTIC. 2006. Getting Paid for Stewardship: An Agricultural Community Water Quality Trading Guide.

EPA. 1996. *NPDES Permit Writer's Manual*, EPA 833-B-96-003.

EPA. 2003. Watershed-Based NPDES Permitting: Rethinking Permitting as Usual. EPA-833-F-03-004. May 2003.

Watershed-Based Permitting:

EPA. 1996. *EPA Watershed Approach Framework*. U.S. Environmental Protection Agency, Washington, DC. http://www.epa.gov/OWOW/watershed/framework.html.

EPA. 2003. Watershed-Based Permitting Case Study: Final Permit, Fact Sheet #1 General Permit for Nitrogen Discharges.

EPA. 2007c. Watershed-based National Pollutant Discharge Elimination System (NPDES) Permitting Technical Guidance August 2007 U.S. Environmental Protection Agency Office of Wastewater Management, Water Permits Division, 1200 Pennsylvania Avenue, NW, Washington, DC 20460, EPA 833-B-07-004.

EPA. 2007d. Watershed-Based Permitting Case Study: Tualatin River Watershed, Oregon Fact Sheet.

EPA. 2007e. Watershed-Based Permitting Case Study: Chesapeake Bay Watershed, Virginia Fact Sheet.

EPA. 2008. Watershed-Based NPDES Permitting, http://cfpub.epa.gov/npdes/wqbasedpermitting/wspermitting.cfm

CHAPTER 6.0

REFERENCE NUTRIENT DISCHARGE PERMITS

This chapter presents a discussion of nutrient discharge permitting issues and some of the special considerations associated with appropriate limits for nitrogen and phosphorus. Example discharge permits with nutrient limits are summarized for reference use. Since special considerations are required for appropriate surface water nutrient discharge limits, the summaries presented in this chapter illustrate the range of effluent limits and variety of permit structures in place for some key nutrient removal facilities.

The discharge permits summarized in this chapter are as follows:

- Clean Water Services of Washington County, Oregon – Durham Plant
- City of Missoula, Montana
- LOTT Alliance Budd Inlet Plant, Olympia, Washington
- Hampton Roads Sanitation District, Virginia
- Stamford Water Pollution Control Authority, Connecticut
- Village of Stamford, New York
- Hillsborough County, Florida – River Oaks Advanced Treatment Plant
- City of Las Vegas, Nevada
- Alexandria Sanitation Authority, Virginia
- Arapahoe County Water and Wastewater Authority, Colorado
- Village of Walton, New York
- Reno and Sparks, Nevada

6.1 Clean Water Services of Washington County, Oregon – Durham Plant

Clean Water Services (CWS) operates four municipal wastewater treatment facilities, each with its own NPDES permit. CWS also has two industrial stormwater permits and is a co-permittee on a Municipal Separate Storm Sewer System (MS4) permit. The Tualatin River is the receiving stream for each of these permitted discharges. Oregon's Department of Environmental Quality (OR DEQ) issued total maximum daily loads (TMDLs) for the Tualatin River for ammonia, phosphorus, temperature, bacteria, and tributary dissolved oxygen (DO).

The CWS permit allows mixing zones and zones of immediate dilution (ZIDs) for surface water discharges. Oregon DEQ issued a single, watershed-based, integrated municipal permit to

CWS. This permit incorporates the NPDES requirements for all four advanced wastewater treatment facilities, its two industrial storm water permits, and its MS4 permit. A significant feature of the integrated permit is its inclusion of provisions for water quality credit trading involving temperature (thermal load), biochemical oxygen demand (BOD), and ammonia (EPA, 2007).

Permit limits for discharges from the Durham plant are seasonal and the plant is required to remove phosphorus and ammonia nitrogen between April and November. During the summer months, the plant must meet an effluent phosphorus concentration of 0.110 mg/L and an effluent ammonia nitrogen concentration of 1 mg/L on a monthly median basis.

The Durham facility located in Tigard, Oregon is operated by CWS of Washington County (District). The plant was designed to operate as a biological phosphorus removal plant in either University of Cape Town (UCT) or A^2O mode, and is typically operated in A^2O. Alum can be added upstream of the primary, secondary, and tertiary treatment processes to meet the seasonal total phosphorus limit.

The biological nutrient removal process follows screening, grit removal, and primary clarification. Lime is added for alkalinity control. Denitrification is practiced to recover alkalinity and oxygen but there is no total nitrogen control requirement in the effluent discharge permit. The tertiary process consists of chemical clarifiers using alum and polymer, followed by sand media filters. Sodium hypochlorite is used for disinfection and sodium bisulfate is used for dechlorination.

Primary sludge is fermented in a two-stage fermenter/thickener, and volatile fatty acids (VFAs) are elutriated and returned to the secondary treatment process. Waste-activated sludge and chemical sludge are thickened using centrifuges. Primary, waste activated and chemical sludges are anaerobically digested and centrifuge dewatered prior to land application. Dewatering centrate is returned to the primary effluent pump station upstream of the aeration basins. Ferric can be added to the anaerobic digester feed for odor and struvite control. The solids process consists of waste solids thickening with a membrane sludge thickener, followed by aerobic digestion, and centrifuge dewatering of digested sludge.

The Durham plant discharges to the Tualitin River and operates under a watershed NPDES discharge permit that includes multiple treatment plants. Discharge permit limits are seasonal and the plant is required to remove phosphorus and ammonia nitrogen between April and November. During the summer months, the plant must meet an effluent phosphorus concentration of 0.110 mg/L and an effluent ammonia nitrogen concentration of 1 mg/L on a monthly median basis.

6.2 City of Missoula, Montana, Wastewater Treatment Plant

The City of Missoula's effluent discharge limits for nitrogen and phosphorus are based on the Voluntary Nutrient Reduction Program (VNRP), a stakeholder driven process that established in-stream water quality targets and defined wasteload allocations for dischargers. The VNRP wasteload allocations were based QUAL-2E modeling of the Clark Fork River with 30Q10 low flows, summer season nutrient controls, and nominal biological nutrient removal treatment for effluent total nitrogen of 10 mg/L and total phosphorus of 1 mg/L.

In October of 1998, the EPA approved the VNRP as the Total Maximum Daily Load (TMDL) for the Clark Fork River. The City of Missoula's discharge permit includes the mass wasteload allocations from the TMDL for total nitrogen of 888.8 lb TN/day and total phosphorus of 88 lb/day. The nutrient limits apply seasonally from June 21 through September 21.

The City of Missoula has operated successfully through two permit cycles with limits based on the VNRP process. The discharge permit includes a re-opener clause that allows the permit to be modified if appropriate for one or more of the following:

- Water Quality Standards: The water quality standards of the receiving waters are modified to require different effluent limits.
- Water Quality Standards are Exceeded: If it is found that water quality standards in the receiving stream are exceeded the effluent limits may be modified.
- TMDL or Wasteload Allocation: TMDL requirements or a wasteload allocation is developed and approved by the state and/or the EPA that modifies the effluent limits.
- Water Quality Management Plan: A revision to the current water quality management plan is approved and adopted which calls for different effluent limits.

6.3 LOTT Alliance Bud Inlet Plant, Olympia, Washington

The Bud Inlet Treatment Plant in Olympia, Washington, is operated by the LOTT Alliance. LOTT is comprised of the contributing jurisdictions of Lacey, Olympia, Tumwater, and Thurston Counties. The plant discharges into Bud Inlet at the south end of Puget Sound. The NPDES permit for Bud Inlet includes limits on Total Inorganic Nitrogen (TIN). Bud Inlet is the only municipal wastewater treatment plant with nitrogen limits that discharges into Puget Sound. There are both chronic and acute mixing zones for the two outalls. The current permit requires LOTT to perform a mixing zone study, as well as a receiving water monitoring study. LOTT studied the quality of Bud Inlet in the late 1990s and found that it has low DO during late summer months, particularly in September (LOTT, 1998). As such, nitrogen limits are part of the NPDES permit. Inner Bud Inlet is listed as impaired for dissolved oxygen (DO), pH, some metals, some organics, and PCBs, and a TMDL is underway. Different seasonal TIN limits are specified. The TIN limit for spring (April and May) and fall (October) is 3 mg/L and 338 lbs/day. The TIN limit for summer (June through September) is 3 mg/L and 288 lbs/day, which approaches the limit of technology for TIN removal. Load limits are on an average monthly basis. The NPDES permit also includes a winter (November through March) total ammonia limit of 22 and 26 mg/L average monthly and 31 and 36 mg/L maximum daily. The two limits specified are for two different outfalls.

6.4 Hampton Roads Sanitation District, Virginia

The Virginia Department of Environmental Quality (VADEQ) has developed a watershed general permit for nutrient discharges to the Chesapeake Bay. Hampton Roads Sanitation District (HRSD) has seven wastewater treatment plants that discharge to surface waters in the James River basin, which flows into Chesapeake Bay. The compliance date for the wasteload allocation is January 1, 2011. A nutrient trading program will be used to meet Virginia's general permit conditions. The associated permits are load based and do not contain concentration limits except for those facilities that are receiving grant funding from the Virginia Water Quality Improvement Fund (WQIF). Each facility has a separate WLA for TN and TP. However, compliance will be judged relative to an annual aggregate loading limit (i.e. bubble limit). The aggregate or bubble limit represents as a sum of these discharged TN and TP loads across the seven facilities.

6.5 Stamford Water Pollution Control Authority – Connecticut

Connecticut developed a General Permit for Nitrogen Discharges that covers all domestic sewage dischargers throughout the state (CTDEP, 2005a). The Connecticut Department of Environmental Protection (CTDEP) and the New York State Department of Environmental Conservations (NYSDEC) developed a TMDL for the Long Island Sound, which was approved by the EPA in 2001. The TMDL left open the option for nitrogen credit trading. The general permit specifies waste load allocations for municipal WWTPs. Each facility's nitrogen limit appears to be based on the total nitrogen allowed from the TMDL allocated by facility flow. The limit is adjusted by an "equivalency factor" that accounts for the "ratio of the unit response of DO to nitrogen in Long Island Sound for each POTW based on the geographic location of the specific POTW's discharge point divided by the unit response of the geographic area with the highest impact." (CTDEP, 2005b) Nitrogen limits decrease from 2006 to 2010 with a final limit to be achieved in 2014.

Stamford WPCA is covered under Connecticut's General Permit for Nitrogen Discharges. The equivalency factor (described above) for Stamford's WWTP is 1.00. The permit specifies annual mass limits for nitrogen. Nitrogen credit trading is based on annual mass loading and credits can be earned by discharging less than permitted nitrogen loads during the course of a calendar year.

6.6 Village of Stamford, New York

The Village of Stamford facility is an activated sludge treatment plant with dual sand filtration of the secondary effluent. The dual sand process uses two Parkson DynaSand™ continuous backwash upflow filters in series. The first filter is approximately two meters in depth and uses coarse sand media. The second filter is approximately one meter in depth and uses fine sand media. A coagulant is added before the first stage filter to precipitate soluble phosphorus and a lamella settler is used to capture solids between stages and improve process throughput. A variety of coagulants have been used in this process, including PASS® (Poly-aluminum-silicate-sulfate), manufactured by Handy Chemical (now Eaglebrook, Inc.). The plant has chlorine disinfection of the effluent. Waste solids are aerobically digested, dewatered in a belt filter press, and landfilled.

The Village of Stamford plant discharges to the New York City watershed, where effluent phosphorus limits are between 1.0 mg/L and 0.2 mg/L depending upon plant flow. The Stamford permit has a monthly average phosphorus limit to 0.20 mg/L, based on a 6-hour composite sample taken twice a month.

From January through August 2005, the average effluent phosphorus reported by the Stamford plant was 0.015 mg/L based on a certified laboratory analysis of the twice monthly 6-hour composite samples. The log normal mean of the twice monthly samples from the summer of 2005 was 0.02 mg/L.

6.7 Hillsborough County, Florida – River Oaks Advanced Treatment Plant

Florida law specifies Total Nitrogen and Total Phosphorus limits for municipal WWTPs discharging into certain water based on design concentrations. Specific limits are set forth for annual average, monthly average, weekly average, and maximum concentrations for a single sample.

Hillsborough County, Florida, operates the 10 mgd River Oaks Advanced Wastewater Treatment Facility (AWWTF) in Tampa. The plant consists of a headworks, including flow equalization and primary treatment; advanced secondary treatment, including nitrification and denitrification; and tertiary filtration, with chemical addition for phosphorus removal. Effluent is aerated and chlorinated before being reused or discharged into Channel, which ultimately leads to Old Tampa Bay.

6.8 City of Las Vegas, Nevada

The City of Las Vegas plant discharges into the Las Vegas Wash, which ultimately flows into Lake Mead and the Colorado River. Seasonal phosphorus and ammonia limits apply to the plant. The mass load allocation to the Las Vegas Wash is shared with two other wastewater plants: Clark County and the City of Henderson. Summer and winter effluent limits for phosphorus at 91 mgd are 0.17 mg/L (126 lbs/day) monthly average. The summer (March through October) effluent ammonia nitrogen limits are 0.48 mg/L (366 lbs/day) and the winter (November to March) limits are 0.56 mg/L (427 lbs/day).

The City of Las Vegas operates a 91-mgd advanced treatment plant that combines an older plant with a relatively new (May 2003) biological nutrient removal facility (BNR). The process includes multiple parallel trains with trickling filters, activated sludge, effluent filters and the new BNR facilities. The older treatment plant consists of trickling filters, nitrification activated sludge, and effluent filtration. Chemical phosphorus removal was practiced at the older plant with chemical addition (ferric) prior to primary clarification. The new BNR facility started operation in May 2003 and treats 30 mgd of the total plant flow. The BNR effluent is combined with the old treatment system prior to effluent filtration.

The relatively new (May 2003) biological nutrient removal facility consists of four 7.5-mgd activated sludge process trains with three anaerobic zones, three anoxic zones, and a complete mixed aerobic zone. The aerobic zone is designed as a racetrack with mixers moving the liquid around the basin. Primary clarification is available with ferric feed as an option, mainly used for odor control at low doses.

The solids processing system consists of gravity thickening of primary sludge, centrifuge thickening of waste activated sludge, anaerobic digestion, dewatering and truck hauling of biosolids.

6.9 Alexandria Sanitation Authority, Virginia

Alexandria, Virginia is a 54-mgd WWTP. Phosphorus removal is accomplished by ferric chloride addition prior to the primary clarifiers, ferric chloride addition following activated sludge ahead of the secondary clarifiers, alum addition prior to chemical clarifiers, and multimedia filtration. Solids are processed by pasteurization, anaerobic digestion and dewatering.

Alexandria's WWTP discharges into Hunting Creek, which ultimately flows into the Potomac River. Hunting Creek is water quality limited for ammonia, fecal coliform, and fish tissue (PCBs). TMDLs for fecal coliform and aquatic life and fish consumption are scheduled to be completed by 2010 and 2014, respectively.

Alexandria has permit limits for ammonia (seasonally adjusted) and total phosphorus. Limits are for both monthly and weekly average concentrations and loads for both parameters.

The one exception is that non-summer ammonia limits are concentrations only. The 50th percentile pH and temperature values for Hunting Creek were used to determine the appropriate acute and chronic ammonia limits (VADEQ, 2004). Summer ammonia limits are driven by the Policy for the Potomac River Embayments. Ammonia limits for non-summer seasons are derived from water quality standards and stream modeling.

VADEQ's internal guidance memorandum recommends dilution factors of 2 and 50 for acute and chronic ammonia toxicity, respectively. Alexandria's permit limits for ammonia were based on the dilution factor of two for acute toxicity, but no dilution factor was used for chronic toxicity due to VADEQ's opinion that dilution was "not applicable for this receiving water body because the discharge is located where the tidal influence is smallest."

Virginia has a special standard for nutrient enriched waters, including Hunting Creek, that requires a monthly average total phosphorus limits of 2 mg/L or less. In addition, year round phosphorus limits are driven by the Policy for the Potomac River Embayments (PPRE). The PPRE monthly average effluent total phosphorus limit is 0.18 mg/L for Alexandria.

6.10 Arapahoe County Water and Wastewater Authority, Colorado

Lone Tree Creek WWTF has a capacity of 2.4 mgd. It serves Arapahoe County, Colorado, in the Denver area. The plant uses a membrane bioreactor (MBR) activated sludge process using ferric chloride to precipitate phosphorus for phosphorus removal. Solids are processed using aerobic digestion and dewatering.

The facility is subject to the *Cherry Creek Reservoir Control Regulation*, Regulation 72, which imposes a total phosphorus concentration limitation of 0.05 mg/l on all dischargers to the reservoir. In addition, the regulation specifies a mass limitation for dischargers of record. For the Lone Tree Creek WWTF, the limitation is currently 402 pounds (lbs) of total phosphorus, which shall be the sum of the monthly phosphorus loads for each direct discharge outfall and land application site calculated for that calendar year.

Total effluent phosphorus log normal average concentration was 40 μg/L in 2003 and 30 μg/L in 2004. Daily results frequently exceed 50 μg/L. Lone Tree is a relatively small plant and the effluent sampling frequency is once per week. There are periods of time where final effluent total phosphorus is 20 μg/L or less, but there are other times when the total phosphorus is much higher.

6.11 Village of Walton, New York

The Village of Walton facility is an activated sludge treatment plant with dual sand filtration of the secondary effluent. The dual sand process uses two Parkson DynaSand™ continuous backwash upflow filters in series. The first filter is approximately 2 meters in depth and uses coarse sand media. The second filter is approximately 1 meter in depth and uses fine sand media. A coagulant is added before the first stage filter to precipitate soluble phosphorus and a lamella settler is used to capture solids between stages and improve process throughput. A variety of coagulants have been used in this process including PASS® (Poly-aluminum-silicate-sulfate), manufactured by Handy Chemical (now Eaglebrook, Inc.). The plant has an influent equalization basin and chlorine disinfection of the effluent. Waste solids are aerobically digested, dewatered in a belt filter press, and landfilled.

The Walton plant discharges to the New York City watershed where effluent phosphorus limits are between 1.0 mg/L and 0.2 mg/L depending upon plant flow. The Walton permit was recently revised to lower the monthly average phosphorus limit to 0.15 mg/L, based on a 24-hour composite sample taken once a week, in order to increase permitted flows to 1.55 mgd. There is also a mass loading limit for phosphorus of 1.95 lbs/day, which was based on the historically permitted flow rate of 1.17 mgd and an effluent limit of 0.2 mg/L.

Walton plant effluent phosphorus data taken weekly from January 9 to August 28, 2005 was reviewed and are shown in Figure 4-13. The log normal mean of the weekly effluent data for 2005 was 0.046 mg/L. The effluent data ranged from 0.01 to 0.49 mg/L in 2005.

6.12 Reno and Sparks, Nevada

The Truckee Meadows Water Reclamation Facility (TMWRF) treats wastewater from the City of Sparks, Nevada, and has a capacity of about 51 mgd. The TMWRF has a flow equalization pond, primary clarifiers, activated sludge with nitrification towers, dual-media filtration, and chlorine disinfection.

Treated effluent is discharged into the Truckee River via Steamboat Creek. Permit conditions include a waste load allocation (WLA) based on a TMDL for the Truckee River. The WLA is based on water quality modeling using DSSAM III (NDEP, 1994). A limited amount of trading of total nitrogen and total phosphorus with two other smaller dischargers is allowed on an annual basis.

6.13 Nutrient Discharge Permit Summary

Tables 6-1 and 6-2 present a summary of the NPDES requirements for the facilities discussed above with respect to averaging periods and nutrients.

Table 6-1. Summary of Reference Nutrient NPDES Permits.

Facility	TN Annual	TN Monthly	TN Weekly	TN Daily	TP Annual	TP Monthly	TP Weekly	TP Daily
Durham (OR)		C (NH_4)				C		
Missoula, MT				L				L
Budd Inlet (Olympia, WA)		C, L (TIN, NH_4)						
HRSD (VA)	L				L			
Stamford, CT	L							
Stamford, NY						C		
Hillsborough County, FL	C	C	C	C	C	C	C	C
Las Vegas, NV		L, C				L, C		
Alexandria, VA		L, C (NH_4)	L, C (NH_4)			L, C	L, C	
Arapahoe County, CO		C (NH_4)	C (NH_4)	C (NH_4)		C		Report
Walton, NY				C				C
Sparks, NV		L		C (NH_4, NO_3)		C, L		

TN = Total Nitrogen; TP = Total Phosphorus; L = Load; C = Concentration; TIN = Total Inorganic Nitrogen

Table 6-2. Summary of Effluent Limits for Reference Nutrient NPDES Permits.

Facility	TN				TP			
	Annual	**Monthly**	**Weekly**	**Daily**	**Annual**	**Monthly**	**Weekly**	**Daily**
Durham (OR)			5.39 lbs/day median NH3-N			0.11 mg/L TP median, May 1-October 31		
Missoula, MT				888.8 lbs/day (June 1-September 30)				88 lbs/day (June 1-September 30)
Budd Inlet (Olympia, WA)		3 mg/L, 375 lbs/day TIN, (April, May, & October); 3 mg/L, 350 lbs/day TIN (June-September); 26/22 mg/L NH4-N (Outfall 001/Outfall 002)		36/31 mg/L NH3-N maximum daily, November-March (Outfall 001/Outfall 002)				
HRSD (VA)	6,000,000 lbs/yr TN allocated between seven WWTPs				582,258 lbs/yr TP allocated between seven WWTPs			
Stamford, CT	1,346 lbs/day TN							
Stamford, NY		2.5 mg/L NH3-N				0.2 mg/L TP		
Hillsborough County, FL	3.0 mg/L	3.75 mg/L	4.5 mg/L	6.0 mg/L	1.0 mg/L	1.25 mg/L	1.5 mg/L	2.0 mg/L
Las Vegas, NV		0.48 mg/L, 366 lbs/day NH3-N (March-October); 0.56 mg/L, 427 lbs/day NH3-N (November-March)				0.17 mg/L, 126 lbs/day		
Alexandria, VA		1.0 mg/L, 204 kg/d NH3-N, (April-October); 8.4 mg/L NH3-N (November-January); 7.4 mg/L NH3-N (February-March)	4.4 mg/L, 899 kg/d NH3-N, April-October; 10.4 mg/L NH3-N November-January; 9.1 mg/L NH3-N (February-March)			0.18 mg/L, 37 kg/d NH3-N	0.27 mg/L, 55 kg/d NH3-N	
Arapahoe County, CO		Monthly average limits vary from 7.5-17 mg/L NH4-N	Daily maximum limits vary from 17-26 mg/L NH4-N					0.05 mg/L
Walton, NY				8.8 mg/L				0.2 mg/L

TN = Total Nitrogen; TP = Total Phosphorus; L = Load; C = Concentration; TIN = Total Inorganic Nitrogen

6.14 References

Connecticut Department of Environmental Protection (CTDEP). 2005a. *Connecticut's Nitrogen Control Program.*

Connecticut Department of Environmental Protection (CTDEP). 2005b. *General Permit for Nitrogen Discharges.*

EPA. 2007. Advanced Wastewater Treatment to Achieve Low Concentration of Phosphorus, EPA 910-R-07-002.

LOTT Alliance, 1998, Budd Inlet Scientific Study Final Report.

Nevada Division of Environmental Protection (NDEP). 1994. Truckee River Final Total Maximum Daily Loads (TMDLs) and Waste Load Allocations (WLAs).

Tri-State Implementation Council. 1998. Nutrient Target Subcommittee, Clark Fork River Voluntary Nutrient Reduction Program (VNRP), August 1998. http://www.tristatecouncil.org/documents/98vnrp_doc.pdf

Virginia Department of Environmental Quality (VADEQ). 2004. *VPDES Permit Program Fact Sheet*, Alexandria Sanitation Authority.

IWA Publishing's authorised EU representative for General Product Safety Regulations is Diane D'Arras, 15 rue Duret, 75116 Paris, France, e-mail: safety@iwap.co.uk.

Printed and bound by CPI Group (UK) Ltd, Croydon, CR0 4YY
17/09/2026
02218798-0014